Astro

C000000150

The Complete Beginners Guide to Discover Stars and Astronomy

(A Very Short Introduction to Astronomy)

Nicole Carlisle

Published By **Andrew Zen**

Nicole Carlisle

Astronomy: The Complete Beginners Guide to Discover Stars and Astronomy (A Very Short Introduction to Astronomy)

ISBN   978-1-77485-712-0

Legal & Disclaimer

The information contained in this ebook is not designed to replace or take the place of any form of medicine or professional medical advice. The information in this ebook has been provided for educational & entertainment purposes only.

The information contained in this book has been compiled from sources deemed reliable, and it is accurate to the best of the Author's knowledge; however, the Author cannot guarantee its accuracy and validity and cannot be held liable for any errors or omissions. Changes are periodically made to this book. You must consult your doctor or get professional medical advice before using any of the suggested remedies, techniques, or information in this book.

Upon using the information contained in this book, you agree to hold harmless the Author from and against any damages, costs, and expenses, including any legal fees potentially resulting from the application of any of the information provided by this guide. This disclaimer applies to any damages or injury caused by the use and application, whether directly or indirectly, of any advice or information presented, whether for breach of contract, tort, negligence, personal injury, criminal intent, or under any other cause of action.

You agree to accept all risks of using the information presented inside this book. You need to consult a professional medical practitioner in order to ensure you are both able and healthy enough to participate in this program.

TABLE OF CONTENTS

Introduction

This book I'd like to guide you on how you can begin your journey in astronomy and become an active observer of the night sky. According to me, it is the very first thing to do is to gaze at the night skies and be fascinated by the universe that surrounds us. Understanding that the earth we live on is just a medium-sized piece of rock and that the solar system itself is medium-sized clump of gravitational particles, and our Milky Way, the galaxy that we reside in is only one of the 2 trillion galaxies that are part of our universe, which help us become more aware of the world surrounding us. The reality that we don't have a unique spot in the universe is a matter of profound implications for philosophy and was the root of the intellectual revolution that's occurred since the time of Nicholas Copernicus and his heliocentric theory of the solar system. Before then it was believed that the Earth was the heart of the universe. anyone who was at the highest point that is the creator also happens to be king of the entire universe. At first, it was discovered there was no sun with other planets that revolve around the Earth instead of the Earth like the other planets orbit around the sun in circular orbits. The lines of space were moved towards the far end to the sun's solar system into space, after it was discovered that the bright dots that appear in the sky are stars that are similar to the sun, and around which planets may also be moving. Next was the need to categorize galaxies as being composed of millions of stars, which greatly influenced our perception of the dimensions of the universe. The reason for the revolution was believed to be due to the discovery that was made by Edwin Hubble, who noticed that galaxies that are distinct are getting further away from us and do so more quicker they get from us. This means that the galaxies used to need to be more close to each other, and from this we can infer the existence of a point in the past when everything was concentrated in a tiny area and had an extremely high density and temperature. It was approximately 13.7 billion years old. The time of this event was described as"the Big Bang, and the scientific validity the theory of Big Bang has been proven numerous times. However, let's return to our expanding universe. One might believe that since all galaxies are moving further away from our planet, then the Earth is at the heart of the universe. This is not the case. It doesn't matter if the galaxies move with respect to space, however space itself expands. This is comparable to baking dough made of raisins. The raw raisins are in close proximity to one another however, during baking the dough expands and distances between them grow for each. Another instance is blowing the balloon into dots. When you inflate it and separating it into dots, the distance between the dots is increased even though none of them is at the center that is the center of the balloon.

In this manner I want to demonstrate how amazing the universe is. It can make me feel overwhelmed. It is a way for us to renew our need to discover the secrets of the universe that we live in as well as the excitement of finding unfamiliar aspects of nature. In the case of this book you've probably felt that way and are aware of the concept I'm referring to.

Chapter 1: The Pleasure Of Exploration.

Next, you must acquire the knowledge. There isn't a solution to this, so you must take time to study the topics that are fascinating to us. It's a good idea to start by reading Popular Science books, which generally are written in a comprehensible language to ensure that we don't get lost in technological aspects related to the matter. This is a position that requires concentration as we move through the various departments, but the pleasure of knowing the process that has thus to date been an abstract concept for our understanding is extremely valuable. In terms of the content of the book is concerned it focuses on the most fundamental questions in cosmology providing the reader with a description of the state of our knowledge regarding the universe. The universe is significantly larger than it was due to the fact that the book was written a long time ago. However, the information contained in it is still relevant and can be used to study the basics of the time and place when the universe was made, what matter is composed of and what are black holes and the challenges of modern physics as a whole. It is interesting to note that the majority of scientists have made the choice to make a choice about their career that was influenced by popular science books. These were the initial stimuli that prompted them to investigate the mysteries of nature.

Purchase an optical instrument

What should I do to begin?

The observation of the sky must begin without the aid of any optical instruments which are not necessary at first. All you require is your vision as well as an atlas and an incline map of the sky. In this way, you will understand the various objects that are visible that are visible in the evening sky. Begin by identifying the brightest stars. You can then look for constellations, which are star clusters that take on distinct shapes such as The constellation known as the Great Bear. These objects will be visible on the rotated Sky Map, which will reveal the location of these objects. Following that is to observe them using binoculars that will give you more details. It is an essential instrument for astronomy because of its mobility, significant magnifying capabilities, and a wide scope of views. Binoculars allow us to see l.a. the moon that will reveal the beautiful craters and mountains, and will be able to see the major planets in the solar system like Mercury, Venus, Mars, Jupiter, and Saturn. Binoculars can also be useful to learn about the very first objects in the night sky that are deep, i.e. objects that are not within the the solar system. They tend to be star clusters and galaxies with bright lights. They are found in the sky atlases.

Which one should you choose?

Next step would be to purchase an instrument. Unfortunately, the scenario isn't straightforward and the selection of a particular model is contingent on several aspects. The most important thing to consider is the object we wish to observe. Our primary focus should be the our solar system which includes planets as well as the moon

The refractor telescope is a good most popular. It's a piece of equipment comprised of a lens system that is compared to an even larger bezel. It was through this lens that Galileo was able to observe Jupiter along with its four biggest moons. With the refractor we can see an image that is inverted and it's ideal for viewing earthly nature and is also smaller and less mobile that is a huge benefit if you are in the city . For observations, you'll go out into the countryside, free of artificial light sources. The lens's minimum diameter for the telescope is believed to be between 50-60 millimeters. Lesser lenses will provide us with more dark images and could not be satisfactory.

If however we are focusing on the deep sky's objects such as clusters, nebulae as well as galaxies then a more effective option is to use an reflector telescope, which is also known as a mirror. The image we receive from it is brighter. The telescope is made up of a massive mirror on the base of the tube which concentrates the in-situ light beam into a narrow one and wraps it up into an additional mirror, which hangs at an angle 45° to on the upper end of the tube. The mirror then directs light towards the eyepiece to which we connect our eye. The image that comes from this telescope will be reversed which makes it not suitable for viewing the scenery. It is believed that the minimum size of the mirror should be thought to be around 130mm.

In conclusion, if live in a metropolitan area and you are concerned about the movement of your telescope and your goals in your observations are objects from the Solar System, the best option would be the refractor. If mobility doesn't matter a lot for you, and your primary concern is deep-sky objects, then it's preferring the spotlight. When purchasing a telescope, it's also essential to expand the scope. The larger the telescope and the greater detail can we be able to see. Magnification is determined by dividing the diameter of the lens with its focal length. In other words If you have a telescope with a focal length 1200mm, and the focal length of the lens is 25mm, then the magnification will be equivalent to 1200mm/25mm or 1200mm/25mm, or. Personally, I would recommend purchasing multiple glasses that allow users to operate at various magnifications. To get a magnification of greater value than what our telescope can offer, we could make use of the Barlow lens. The most commonly available available are those that have 2x, 3x or 5x multipliers.

What are the most important things to be aware of when purchasing an instrument?

One of the most crucial aspects that a telescope must have is its location, i.e. the tripod that it is mounted on.

The tube is put in. It is the first step. must be stable so that the image displayed in the eyepiece isn't distorted and compromise the quality of the observations. It is possible to divide the telescope into two major groups: azimuth and parallax. The former lets us operate the telescope upwards and down, and revolve around its axis. however, they don't allow for photos of celestial objects. Parallax assemblies, on the other hand are designed according to the geographical directions and the axis of rotation of Earth, which means that the telescope will always be aligned in one direction that is ideal for photographing with long exposure durations. Another thing to check out for is the kind of lens that is that the kit comes with. planning to purchase. It is vital that we purchase a telescope with a good quality.

Equipped with a set glasses, at minimum two glasses, an assembly, and an instrument that is utilized to

the telescope. A crucial element is the filter, with which we are able to

look for the sun or observe the full moon or the sun. If you are looking through sales ads look for the full moon or sun. You will surely see the so-called GOTO system that lets you automatically adjust the telescope once you have entered your name for the object. This is an extremely convenient option that can cut down the search time from a couple of hours to just a few seconds. I believe this is a rebuke to becoming a victim of an arrangement, as without it, we'd be in complete limbo. Furthermore, I believe observation, in the process of locating something within the night sky which is interesting to us is incredibly enjoyable and satisfying.

3. What should you look for at night?
The process of determining the coordinates in the sky, or, the distances in angular terms
Before we start to talk about what we could observe in the sky, I'd like details on how

Angular distances are. They could prove essential. In the atlas of the sky we can see that two objects are approximately 10 degrees from each other. This signifies that distances between two bodies within the Great Circle or the imaginary firmament-sphere is only 10 degrees. Of course, many people will not be able to test this using instruments, such as sextans since for amateur sky observations more appropriate, quicker and simpler methods exist. This method is to measure angles in the sky with... hands. The existence of a few guidelines to follow will enable us to calculate these numbers easily. These are the rules:

The length of the little finger when extended to the width from the palm is approximately 1 degree

The size of the middle fingers of 5 degrees

The angle of the fist that is clenched has a 10 degree angle.

The distance between the biggest index and the smaller fingers is 15 degrees

The distance between the thumb's maximum and the small finger is 25 degrees.

Overview of Astronomical objects

When we've got the telescope of our dreams It's time to begin planning observations. To begin this process, one must first know the objects we can spot within the evening sky, so that the image that is seen by the eyepiece more than a assortment of sparkling dots. A brief overview of the stars and astronomical objects that are visible from our backyard, including the solar system:

1. The Moon The first thing I'd suggest you rotate your telescope towards would be the moon. The reason is that it is the biggest and brightest star visible at night, meaning we are able to locate it. It's a tiny accomplishment at the beginning dry riverbeds that stretch across hundreds of kilometers with mountain chains. When I first saw the moon with the telescope, it was hard to be more amazed at how fascinating the location was. I would also suggest that you begin your journey into astrophotography simply by looking at the moon, which displays its best through the lens, and the images you take are sure to bring you a lot of pleasure. In addition, it is

worth noting how to go about it and the equipment you'll require. A smartphone that has an integrated camera is sufficient for the initial step. Attach your phone to the eye-opener, and adjust it as needed this could cause minor difficulties, but isn't difficult. One way to avoid this dilemma is to purchase an appropriate tripod for the phone, which allows it to be permanently connected with the lens.

If we wish our photographs to be better quality you can get professional cameras by mounting it on the telescope by using an adapter specifically designed for the image below.

Below, I'll also share one moon-related photo that I was able to capture using my smartphone.

The final thing about the moon that I'd like to highlight are lunar filters placed on the lens. They offer a greater quantity of details that can be seen through the telescope , and safeguard your eyes during times when the moon's fully illuminated and shines extremely bright.

2. The Sun is a star that we can observe. Sun using a proper filter is required, as I've

mentioned before. Once we have this filter then we can look at our closest star that, despite its appearance not need to be boring in any way. If the telescope is directed towards the sun, we'll be able see distinctive areas on its surface resultant from a decrease in temperature over the area. One interesting thing to look for is an event known as transit i.e. traversing an outer solar shell of one planet. If you're fortunate or have a budget, you may be able to visit the spot in which the total solar eclipse can be seen, that is certain to make for a memorable experience.

3. Jupiter is the top planet that can be observed among the planets in the solar system. It is a gas giant and is the biggest celestial body that we have in our solar system. It is a sight to behold with its distinctive Equatorial stripes. And in telescopes that have high magnification , we are able to observe the Great Red Spot. Notable are also its four largest moons which are Ganymede, Io, Europa and Callisto that constantly alter their positions with respect the shield of Jupiter.

4. Saturn is the second largest world in the Solar System, which we can see easily for Utilizing a tiny telescope with a small size. A distinctive ring of rock particles and ice extends around Saturn. With the help of optical instruments that have more capabilities, we'll be able to observe the known as the Cassini gap which is the largest of visible ring fractures.

5. Venus is also known by the name The Morning Star, because it's most visible in the sky in early morning when all stars are visible. When viewed through a telescope we see it as a glowing dot or a sickle based on its the phase.

6. Mercury is the planet closest to the sun and largest satellite in the solar system. Due to its proximity our star, it's generally obscured and not as appealing an object of astronomical significance.

7. Mars - the planet for which we have the greatest hopes for colonization in space. Manned missions to Mars are in the works by Elon Musk's firm SpaceX from the beginning of the 2030s. The telescope is an orange ball on which we can see outline of the white ice caps that are visible on the

poles. If we are lucky we may even be able to discern the shape of Sand storms.

8. Uranus along with Neptune have already become a greater task for the beginner observer, but with time, you will be able to spot these two stars on the night skies. Uranus appears as a greenish shield , while Neptune appears to be a slight blue shield.

9. Asteroids are rocks that are located within their orbits around Mars as well as Jupiter. The largest is Ceres that was first discovered by 1801. The brightest ones aren't difficult to spot and knowing the current location of an asteroid ought to be simple to determine.

10. Star clusters - they're typically divided into two types: open and spherical clusters. They are gravity-driven clumps made up of hundreds or tens of stars that are loosely connected. They usually take rather different shapes. They are very young objects, usually featuring CA. Tens of millions of years. The most well-known open-cluster can be found in one called the Pleiades (M45) within the constellation Bull. M45

Globular clusters are gravitational groups of stars that are closely connected and whose age is comparable to the age in the Universe. When we look at these globular clusters using binoculars, we usually are capable of seeing a vague area, however when our telescopes are pointed towards this location with a large magnification, it is possible to see an image that is composed of thousands of tiny dots. A single of the renowned types of structures is the large globular cluster that is located in the constellation Hercules(M13).

M13

11. Nebulae These are enormous cloud of dust in the form of stars that cover the area of many light-years. We can distinguish two types of Nebulae. The first is planetary nebulae that result from the explosion of stars A prime instance of a nebula like this is the ring found in the constellation Lute(M57).

M57

The other type is diffusion nebulae. These are huge cloud of dust and gas which take on shape that is irregular. One of the most famous Nebulae that belong to this class includes the Great Orion Nebula(M42).

M42

It is interesting to note that in 2013 a nebula made mostly of ethyl ethanol was discovered. Its diameter is over 1,000 times that in the solar system as well as the quantity of ethanol in it is believed to be sufficient for humans to last for up to 1000 many times more than expected life span of the universe which is measured by millions of years.

12. Galaxies are massive clusters of millions of star clusters that have been placed into distinct structures. We recognize spiral, elliptical and lenticular galaxies, as well as irregular galaxies. The solar system is situated within a spiral galaxy known as The Milky Way, which can be observed through the evening sky in a broad strip that is slightly brighter than the other sky. Alongside being able to see the Milky Way, we can see other galaxies as well that are close ours, and therefore the brightest is known as the Andromeda galaxy, which is located 2.5 million light years away, making

it the most distant object of the night skies that is visible by the naked eye.

Milky Way

Andromeda Galaxy

13. Comets are celestial bodies that are comprised of ammonia, ice methane, carbon dioxide. They can be classified into regular ones i.e. ones that, when they travel through an elliptical orbit return time and the sun's surface and non-periodic ones which are determined by a hyperbola or a parabola. As the comet moves closer to towards the sun's surface, it develops two braids because of the interaction with solar winds. dust and gas. The gas plume is blue and is located in the opposite direction from the sun. The white dust, however is curled to follow the comet's movements. The most well-known is Halley's Comet, which is seen returning to the sun each about 75 years. The oldest evidence of the comet date back to 613 BC and its coming back in 2061.

14. Meteor swarms , Finally, I'd like to highlight the phenomenon known as "Shooting stars". This happens when a rock granule gets thrown into the Earth's atmosphere and then burns, leaving the

appearance of a white streak. Shooting stars can be seen at night almost all the time, but there are certain times of greater activity, like meteor swarms like those that occur during the month of August known as Perseids. The amount of shooting stars observed can increase to several thousands per hour. The observations of shooting stars typically take place with an eye that is not armed or, less frequently, with binoculars or telescopes.

As you can see that you are surrounded by a myriad of options and things to look at will likely never end and astronomy is an occupation that will last for a lifetime. The night sky that seems to be a fairly homogeneous creation made up of glowing dots, conceals an amazing array of shapes waiting patiently for you to discover them. As you gain astronomical proficiency and you improve your astronomical skills, you can use your telescope's lens towards more detectable objects, allowing the continuous growth.

Chapter 2: What Do You Get Ready For Observation?

Selecting a location of observation

In this chapter I will go over step-by-step how to prepare yourself for observations and how to carry them out. After this chapter is finished you'll be able to take the telescope out and start your journey in Astronomy.

The first step is to determine the site of the observation this is a vital aspect. The observatory must be as clear of any source of light which can hinder the appearance of the objects within the sky at night. Another factor in deciding on the best location for observation is the absence of buildings with high ceilings that completely block a specific portion of the sky. If you're located in the countryside, you are likely to find a spot close to your home, but when you live in a city, you might have to look out from the area you reside. However, even in big urban agglomerations, you can still be found in areas with dark skies, however, there are smaller numbers.

What are the best times to make observations?

After you've decided the location of your observatory It is now time to decide on the date for your observations and the direction you'll direct your telescope towards. You must be aware that beginning observations without having a clear plan is completely unwise. There's a lot of excitement about possibilities, but chances of discovering something of interest is low. Start by figuring out when the ideal conditions will be present and then mark the date in your schedule. What are these ideal conditions? Most importantly, it is there are no clouds that makes it impossible to determine the right direction of the moon's phase as well as the conditions. Overcast is a fairly regular phenomenon that can not be forecasted for more than a couple of days, so don't try to schedule your observations with one month's notice because it's not logical. Moon phase however can be verified as a aspect, and we are able to observe it without prejudice. If you're looking to see your view of the Deep Sky, choose the time that is closest up to New Year, during which the moon appears to be the darkest for anyone who is watching it on Earth. It is also possible to see at what time the moon is located on the opposite side of the globe , and remains hidden to the naked eye. It is possible to watch for a long time in the event that the moon is at its peak. Of the course, if your goal is to look at the moon each phase is perfect for this. Another factor to consider is weather, which could interfere with our plans. Therefore, be sure to ensure that no rain or any other rainfall will be forecast in this forecast Day and also that it's not too frigid.

The process of observation planning

After you've established the place and the time of the observation It is important to take care to plan your journey. For the first time I suggest reading the no-cost Stellarium programthat lets you watch an animation that shows the sky's nighttime skies any moment during the day from any place on Earth. At this point we can choose the stars that are the most visible at night, and create the plan of observation. In this regard it is recommended to create a notebook that we can use to note down all celestial bodies that are interesting to us during the night. If you are using the

Stellarium application, it is advisable to carry an atlas of the sky on hand to be able to access all the information we are looking for. The primary criterion used to select an object for observation is its brightness also known as Stellar size which is also called magnitudes(mag). It is a measure that defines how bright celestial objects with smaller values indicate greater brightness, and consequently visibility. The full moon is an object with a stellar magnitude (12 mag) as well as the Sun is brighter than the Sun with a luminosity of -26mag. When choosing celestial bodies for observation make sure you pick ones with the lowest magnitude, since objects with the size of a star equal to 10 will be difficult to locate on the dark sky. To find out more, look up Messier's catalogue (M) which includes a handful of exceptions that includes the most beautiful objects visible in the night sky. Alongside Messier's catalog, there's an array of stunning astronomical objects known as NGC.

Both catalogs can be found at the conclusion in your Sky Atlas. When you are drafting your observation plans, be sure that on the top of your list are the stars which are the first to reach the Horizon. This is why I'd also recommend not to overly expect anything from your telescope. Unfortunately, because of the limitations of amateur optical equipment we won't get images that we can view through the web. They're made by massive ground-based or space-based telescopes which gather data for a long period of time. The image itself is the product from computer-generated processing. If you're looking to get the highest quality images, I suggest the use of a computer program known as RegiStax that employs the technique of stacking. It involves loading several images of the same subject or short videos into RegiStax. The computer will then choose the most appealing frames and blends them together to create the best quality image that is possible. It's important to remember that the majority of objects that you can see with the telescope will be visible in grey shades and won't be able to be able to see the depth of the shades of space. When you've got a thorough observation plan, there's only one choice: join them.

However, before doing you do that, I'd suggest the following important aspects. The first is to place the telescope in the open for an hour before night sky survey starts. The reason this is done is due to the difference in temperature that is created between the air inside the tube of the telescope and the air outside. This could cause an image waveform that significantly reduces the quality of the view. The same effect can be obtained by looking at the sky through a glass, something I strongly recommend against. The other thing to consider is the adjusting of the eyes' brightness. The complete adjustment of our eyes to darkness happens after an hour. This allows us to see greater details in the sky, which is exactly what we want. It is imperative that care take to make sure our eyes aren't brought in contact with bright lighting for at least an hour prior to observation. Many people use sunglasses to do this, or even completely cover the eyes by which they are looking. It is possible to ask when we have to light for instance, an image that rotates in the sky in the course of observation. Well, there's away. It's as simple as using an LED that emits red light. This can reduce the eye's ability to adjust to darkness. If you don't have the flashlight, you could decorate the mirror of a regular lamp with red nail polish. The last question is what you should bring along to watch the sky. Along with an telescope and a pair of glasses, carry a rotating map of the sky, Sky atlas, a notebook with an observation plan, a flashlight equipped with the red light, binoculars, hot tea and warm clothing.

5. How do you conduct observation?

How can you locate stars in the night sky?

In this stage we can begin to outline the procedure for making observations. It starts by identifying the object we would like to see and finding that object in the night sky. This is made easier by the moving Sky Map, which shows the locations and position of star clusters in the evening sky dependent on the time of the year and the day of the week. If you want a map that is rotating the sky that can show us the picture we see above our heads the moving portion of the map has to be rotated to ensure that the day you are looking at corresponds to the observation time. The easiest way to identify specific celestial objects that are visible in the night sky can be done to use an interactive application that you can download to your smartphone, which according to where you position the camera gives you the exact view of sky. In my opinion I would recommend SkyView, which is available for both SkyView application available for iOS or Star Walk for Android. It is a great solution for people who are just beginning, as it will give you a greater understanding of the structure of the night sky. After you've identified the object of your interest I suggest looking at it with a greater depth. the object using binoculars which can make us feel more confident about being able to locate him in a crowd of stars. It's now time to point the telescope at this point. It is important to note that initially, finding deep sky objects isn't easy and can require a couple of minutes, but trust me when I say that the feeling of finding something that is thousands of light years away is incredible. Don't let yourself be dismayed if you are having difficulties with it. It's all about patience, and if you persevere, you'll see improvement in your ability to navigate through the sky at night. Okay, so how can you use the telescope to locate the object? The first step is to point at the Seeker which is a small set of binoculars mounted on the telescope, the location where the celestial body is. Be sure to set the Finder to the correct calibration. This can be done by pointing the telescope at an undetermined location, and then turning the knobs the location of the seeker to ensure that the image you see in the eyepiece is similar to what we see through the seeker. After steering the telescope towards the correct spot on the sky it is time to focus on the image using the knob located next on the lens. Then you can begin searching towards your goal. Of course, it's impossible to always direct your telescope in the right position, so we employ tiny inches i.e. very tiny changes in one direction to adjust the telescope's place. It is also crucial to begin your observations with the eyepiece that has the lowest magnification that will give us the widest field of view. When you've found something I would suggest spending a few minutes to look at it and then write notes in your notebook.

Messier's Marathon

If you are able to improve your astronomical skills I would encourage you to participate at the Messier Marathon that is an all-night sky observatory that aims to spot as many objects as possible using Messier's catalogue. This is an excellent challenge to endurance and understanding of the night sky. Typically, the observations begin from the western horizon because this is where the stars appear the most quickly. They then move gradually to the eastern portion of the sky. Finally, it ends the marathon with the initial sunlight. The most obvious benefit of such observation is the fact that they expand our understanding about the sky at night, and also encourage us to look at all the objects that Messier's catalog has and not just the most popular objects. The main drawback is the time we're able to spend with an individual celestial body since when we look at around 100 astronomical objects, there are only just a few minutes to spend on one. However, the marathon can be a great occasion to get together with astronomers and to experience the astronomical objects.

Chapter 3: Stars

Stars are the huge, bright glowing ball of gas (mainly hydrogen and the helium) they play a major role in the creation of the vast most chemical elements (from which we humans are made from) I love to claim that I'm made from stars dust (because we are made from elements that originated from stars, after they had died).

They are among the most complicated celestial bodies that exist in the universe. At the beginning of astrophysics, scientists are not aware of the fact that this is where the stars obtain their energy until 1930s when the famous scientist Hans Bethe found out that nuclear fusion was possible and was the an energy source for stars.

There are many kinds of stars in the universe, however only a few of them consist of Red gigantic, Blue giant, Supergiant.

The process of formation of stars is truly amazing to watch this process superior to watching any movie.

It takes millions of years to develop into the bright, luminous star this process starts with the formation of massive clouds of molecules, atoms, dust and gas. The the molecules that are present in clouds are H2.

The molecular cloud is just a few degrees higher than absolute zero, but once the cloud begins to break down, it will begin to warm up, the outside of the cloud is 10degk, while the inner one cloud is 300degk(room temperature) at the point that the core will exceed 10,000 Au(Astronomical unit-distance from sun and earth) It is referred to as pre-star(stage prior to becoming a star) within the in the next 50,000 years, its core will contract and after 50,000 years, the system will develop into a disc around its core. The access material will be blown away into space. From the poles(as the earth also has) specifically axis, in which stars spin around. You will see a fountain-like structure in which this extra material is released. Jets are the structures that form which are the random movement of gas and dust we discussed earlier, when combined with the system's shrinking as the pre-stellar central core is formed it will cause the whole system to spin. This results in a flat disk to develop around the pre-stellar core. This is similar to process that creates the shape of a flat disk around an ice skater who is spinning. If the skater wasn't turning then the dress wouldn't be a smooth disk surrounding her however, it would instead hang from her sides. The jets that are at the poles are created to help keep the balance of the system. This system has been known as proto-star. This means it is in its beginning of the process of becoming a true star.

The disk is vital for a proto-stars to develop to become a star of the proper size. The disk is composed mainly of gas. It is rotated with the disk as it slowly creeps closer to that surface. If the gas is closer to the stars it falls on it due to the force of gravity and the star expands. This process of growth is known as an Acretion (the process in which one object collects the mass of an object) process, and the star is believed to absorb (accumulate) material from its disk.

Over the course that spans hundreds of thousands years the star has increased in both density and size for the central region to trigger the nuclear reaction, which makes the star shine similar to the Sun. At this moment, the star is classified as a T-tauri star (a extremely young star, we'll discuss it in the following chapter) This marks the very first occasion the star is visible visually. The star will eventually stop in accreting matter out of the disk, however, the remaining material surrounding the star remains in a disk-like form. The disk is no longer serving its purpose of feeding to the star matter and making the star expand. Instead the disk has become simply a moving circular plane of materialthat will gradually begin to become clumped and eventually become orbital around the star. These tiny clumps, formed by the remnants of material that remained from the creation of the star will create new planets. That means that the planets that we have in the solar system we live in are composed out of leftover material left that was created by the Sun! This is the reason why all planets in our solar system can be on the same plane during the process of forming of the sun. As the solar system is formed. So, stop for a moment and consider how amazing it was going on when the stars from silence to violent (at the moment of ejection) and finally it formed the whole solar system, not just itself, but over the following 10 billion years, the star will begin to burn the nuclear fuel that is in its core and emit energy through the sun's radiation.

I am always looking at the night sky and consider the stars that are in the sky that are visible to us . There are billions of stars which aren't visible to our eyes and again, consider how far they aren't accessible to us. A lot of people do not even know there is much to know about stars, that's why I'm trying to write as much as I can in a simple way.

We have seen how stars are created. it's time to discover who the massive ball of fire will is going to die, because we all know that if something or someone was to be born in the universe, it will have to go to the grave regardless of how strong or huge it may be.

Before we know the process by which stars die, we need to know the inner workings of the stars. When the star begins its nuclear fusion, one powerful force working in opposition to it, and that is the gravitational force , which is trying to break up the entire the star's core. now you understand that there are two forces that are working in opposition to one another within the stars. One is nuclear the fusion (I refer to it as"the force that lives") and the other gravity force (death force) It's time to look at the way that stars die.

The tiniest stars capable of this feat are known as red dwarfs because they're colored red and are small in dimensions. They could be as tiny as one tenth of the size of the sun's mass. They are the ultimate economical cars in the universe's highways capable of consuming hydrogen fuel for billions of years.

Medium-sized stars, such as our sun, exhaust their hydrogen reserves in just one billion years. This is enough time for small creatures to make their way to the surface of a water-filled planet and ask questions. If stars like our sun go out of their orbit, they transform into the most bizarre slow-motion horror and eventually reveal their oxygen and carbon cores, and leaving behind a sparkling Nebula.

If the stars of the night stop fusion of hydrogen inside their cores initially nothing unusual occurs. The hydrogen fusion results in helium that because of the extreme pressures can be ignited during its own the fusion process, which is being surrounded by a layer burning hydrogen. In the end, as you may have already guessed, the helium inside the core disappears. The fusion of helium leads to oxygen and carbon. Then, it is fused with silicon, leaving behind magnesium.

The final phase of a star's career the pace is a bit chaotic. There's a core comprised of iron and nickel that can reach a temperature of more than a hundred million degrees. This core is protected by hell's seven-layer bean dip, which is made up of shells that fuse silicon, magnesium oxygen, carbon, hydrogen, and helium. As soon as the iron core has formed the clock begins to tick. Within about one hour, the spectacle will end.

The issue lies in the fact that when you combine elements that weigh less than iron, you'll get only a tiny amount of energy absorbed in the process. It's a good thing that stars are in the mix. Once you start to fuse iron with elements heavier than iron it loses energy during the process. This creates an unstable situation. Gravity continues to for the star to be pulled in however there isn't an explosion of energy that can ensure it stays in place.

The iron core simply gets squeezed and squeezes down to such incredible levels that electrons are pushed (push) into protons, which transform the protons into neutrons. Within a couple of minutes, the whole core shrinks to a gigantic ball of neutrons roughly what is the equivalent of city trillions of times dense--a neutron star. This neutron star is able to temporarily stop the collapse of the star around it. However, all that gas and the flaming layers, fall at a significant fraction in the rate of light. They strike the neutron star, and bounce off.

BOOM (Supernova)

If a supernova does occur, it's one of the most intense and brightest cosmic events known to man. One single explosion of a supernova could outdo hundreds of millions of stars. In just a moment, the supernova can produce more energy than the sun in its entire lifetime.

I have mentioned a few kinds of stars, such as the t-tauri stars and neutron star. Let's talk about them. T-Tauri star are the type of variable stars less than 10 millions of years old. They are discovered in the vicinity of molecular clouds and are distinguished through their optical variability and distinct Chromoshperic lines. You're now able to know what a neutron-star is? The answer is that it's an exploding core from a supergiant star that was massive with an overall mass of between 10 to 25 solar masses and possibly greater if the star was particularly metal-rich.

We've talked about a variety of topics the star, such as what it is what is it? How did it get its name? and dies? Most readers are aware that the sun is composed of many layers. It's time to look at the entire spectrum of layers before we get started. I'd like to reveal a fascinating information: if you determine the amount of time a photon (light particle) require to travel from the sun to earth, the answer will be approximately 8 minutes. This means that we can conclude that the photon arriving at earth is 8-9 minutes old. The it is not the case. photo could be 10 to 100 thousand years old. then you're all wondering how this is possible, but the main reason for the fineness of a photon is its own layers, and the number of photons in the star increases the difficulty and make it hard for photons to emerge. It's now time to conduct an autopsy on the star. The layers that are in the middle are those of Core, Radiative Zone and Convection Zone. The outer layers comprise the Photosphere as well as the Chromosphere as well as The Transition Region and the Corona.

1. Core: It's the central location from which all nuclear fusion processes occurs. It's akin to the engines of stars, from which stars obtain energy. it's the same location that star collapses if the stars ' fuel is exhausted and gravitational force wins.

2. Radiative Zone The term "radiative zone" or radiative area, is an outer layer of a star's interior, where energy is mostly transported towards the outside by thermal conduction and radiative diffusion instead of convection. The energy moves throughout the zone as a result electromagnetic radiation, which is referred to as photons.

3. Convection Zone Convection Zone the outermost layer of the interior. It extends

from the depth of 200,000 km all the way towards the visible portion that is the Sun. Convection is the method of transporting energy within this zone. The convection zone's surface zone is the place where the light (photons) is produced.

4. Photosphere - The photosphere the deepest layer of Sun which we can view directly. It extends from the surface visible in the middle of our solar disc up to around 250 miles (400 kilometers) above. The temperature of the photosphere ranges from 6500K at bottom to around 4000 temperatures at the highest (11,000 as high as 6700 F 6200 degrees F, and 6200 to 3700 degrees Celsius). The majority of the photosphere coated by the granulation.

5. Chromosphere: The chromosphere is part of the Sun located between 250 miles (400 kilometers) to 1300 miles (2100 km) above the surface of the Sun (the photosphere). The temperature of the chromosphere ranges from 4000 K near the lowest (the known as the temperatures minimum) to 8000 K in the highest (6700 to 14,000 F 3700 and 7700 degree C) and, therefore, within this layer (and more advanced layers) it is

actually more hot if you move farther away from Sun but not so in the lower layers which get hotter if you move closer to the central region in the Sun.

6. Transient region: The region of transition is a slender (60 miles or 100 kilometers) layer that lies between the chromosphere as well as the corona, where temperatures rise quickly from around 8000 up to 500 000 K (14,000 to 9000 degrees F, 7700 up to 500,000 degree C).

7. Corona The corona is the outermost layer on the Sun that begins at around 1300 miles (2100 km) over the solar surface (the photosphere). The temperature of the corona can reach at 500,000 kilometers (900,000 degrees F 500,000 degrees C) or greater, all the way to just a few millions of K. The corona is not visible through the naked eye, except when there is an eclipse of the total sun or through the use of coronagraphs. The corona doesn't have any upper limits. Source: National Solar Observatory.

All the sun's layers.

It's the season for imagination take a look at the stars we can see during the time of daylight (sun) in addition to the ones that we see during the evening. What is so amazing is most people believe that twinkling stars at night are be seen only at night, but the truth is that they are always visible in the sky, however because of the enormous solar brightness, they are such a dim light that we are unable to see them through our naked eyes.

Another thing that fascinates me is that it's not required for us to know that the stars that can be seen at night are alive. Does that not strange? light is the most well-known thing in the universe, but the distance between the earth and the night-time twinkling star is huge enough that light emitted by stars takes billions of years of light to reach earth before coming in your sight (I swear by the information I gave you this moment is accurate) you are not interested to you? That's how big the universe is that even light has to take 4.24 days to get to us from our nearest star Proxima Centauri. If you don't know how many light years is, it's just multiplying the c (speed of light) by seconds. In one year. You can determine the distance in the human sphere, but for the cosmic sphere it's nothing since the area of the known universe is around 90 billion light years (multiply the speed of light by the total number of seconds in nine billion years) I'm sure it's excessive, but it's an actual story, and that's why I consider astronomy to be the most real science fiction.

Stars are so intricate that physicists devote their entire lives studying about stars, watching it and trying to resolve the mysteries of stars that remain unsolved. among the top well-known scientists who have devoted his entire life to stars is Subrahmanyan Chandrasekhar the an Indian American astrophysicist who found the Chandrasekhar limit. It's the amount of mass that the pressure of electron degeneracy in the star's core is not sufficient to counterbalance the star's gravity-dependent self-attraction.

It is believed that the Chandrasekhar limitation is currently considered as being approximately 1.4 times the solar mass; any white dwarf having less than this amount will remain as a white dwarf for the rest of its life however, a star which exceeds this amount is likely to die by the most violent of explosions, the supernovae

They also create magnetic fields around them. It is referred to in the form of Stellar Magnetic Field I call it SMF in its short form. Let's find out what SMF is? This is the term used to describe a field of magnetic energy that is created by the movement of conductive plasma within the star. The motion is caused by convection that is a kind of energy transportation that involves physical movement of materials. A magnetic field that is localized is a force that exerts on the plasma, thereby increasing the pressure, but without a similar increase in density. In the end, the area of magnetization increases in comparison to the rest of the plasma, and eventually it reaches the photosphere of the star. This causes starspots (discuss further) in the plasma's surface as well as the similar process in Coronal Loops (discuss further).

We now understand how stars generate magnetic fields and we have some new terms like coronal loop and starspot. Let's look at what they mean? Starspot are the result of a temporary phenomenon in the photosphere of the star which appear as dark spots than surrounding areas. They are areas with a lower temperature due to large amounts in magnetic flux which hinder convection. The Sun also contains spots along its surfaces. These spots are so massive that the entire earth could be encapsulated in one dark spot but they aren't visible from Earth because sun's light intensity covers the dark spots. What is coronal loops? They are massive magnetic field loops that begin and concluding on the Star's apparent surface (photosphere) that extend onto the sun's atmosphere (corona). Ionized gas that is hot and glowing (plasma) caught in the loops allows them to be seen.

The magnitude in the magnetic field fluctuates according to the size and composition of the star. Likewise, the magnitude of surface activity on magnetic surfaces is dependent on the rate at which the star is rotation. The surface activity causes coronal loops, starspots, and starspots. Stellar flares (sudden flash of higher brightness).

Stars that are young and rapidly rotating typically exhibit significant levels of activity on their surfaces due to the magnetic field they create. The magnetic field may act on the star's stellar winds and act as a brake that slowly slow down the speed of rotation over time. So older stars like the Sun are slower in their rate of rotation as well as less surface activity. The activity levels of slow moving stars can fluctuate in a cycle and may even cease to function for long periods of duration. In the Maunder minimum (is the name that is used to describe the time period from 1645 until 1715 when sunspots were extremely uncommon, as observed by solar observators) For instance during the Maunder minimum, the Sun was undergoing a 70-year stretch without sunspot activity.

The energy created by stars, which is a result that results from the nuclear process, emits into space in the form of electromagnetic radiation as well as particle radiation. The radiation that is emitted from stars is reflected as the stellar wind that flows out from the outer layers in electro-charged protons as well as beta and alpha particles. A continuous stream of non-massless neutrinos is emitted directly from the core of the star.

The creation of energy at the heart of the star is the reason why stars shine so brightly. Every when two or more nuclei of atomic nature fuse together to create an atomic nucleus that is the heavier element Gamma rays escape from the products of fusion. The energy is transformed into different forms of electromagnetic energy that are of lower frequencies, such as visible light at the point it is released into the outer layers.

The colour of a star's surface as measured by the highest frequency of visible light is determined by its temperature. outer layer of the star which includes its photosphere. Apart from the visible spectrum, starlights release electromagnetic radiation in forms which are inaccessible to our eyes. Actually the spectrum of stellar electromagnetic radiation covers the entire electromagnetic spectrum including the most extensive radio waves' wavelengths through visible light, infrared and ultraviolet, and the shortest X-rays and the gamma radiation. From the point of view of the all the energy that is released by a star, all of the components of the electromagnetic radiation of stars are important, however all frequencies offer an understanding of the physics of the star.

Utilizing the stellar spectrum, Astronomers can measure the temperature of the surface, its gravity and metallicity (is the quantity of elements found in an object which weigh more than hydrogen or the helium.) and the rotational speed of the star. In the event that the location of the stars is determined via measurements such as the parallax, the brightness of the star is calculated. The radius, mass as well as the surface gravity and the period of rotation can be calculated based on stellar models. (Mass can be determined for stars in the binary system by measuring orbital velocity and distances. Microlensing with gravitational force (discuss in another chapters) can be employed to determine the mass of one star. By using this information, the astronomers are able to determine the age of stars.

We've already learned the process of dying stars and the formation of neutron stars, but this isn't enough to comprehend everything. We need to discover more about the collapse of star.

The final stage of stars takes place when a huge star starts producing iron. Because iron nuclei are more tightly bound than other heavier nuclei, any fusion that goes beyond iron doesn't result in the energy in a net way as a result of which star's cores begin to shrink.

When the core of a star shrinks and the intensity of radiation emanating from its surface increases and creates such a pressure (discuss in a future post) within the layer made of gas. It can remove the layers creating an planetary Nebula. If the remaining space after the atmosphere's outer layer has been removed is less than 1.4 millimeters, then it will shrink to a tiny object that is about equal to Earth which is known as white dwarf. White dwarfs aren't massive enough that allows further compression of gravity to occur. The electron-degenerate material inside the white dwarf is not plasma anymore. At some point, white dwarfs disappear into black dwarfs after a long time.

In massive stars where fusion continues, it will continue till the core of iron is to the point (more than 1.4 M) that it is unable to be able to support its own weight. The core will fall apart as electrons are pushed into protons, creating neutrinos, neutrons, and gamma rays during the form of a flash that includes electron capture as well as an inverse beta decay. The shockwave created by this sudden collapse causes the remainder part of the stars to explode into an explosion known as a supernova. Supernovae get so bright they can briefly eclipse the galaxy that is home to the star. If they happen in the Milky Way, supernovae have previously have been observed by non-naked observer for their "new stars" when none existed before.

A supernova explosion shatters the outer layers of the star leaving behind a remnant like The Crab Nebula. The center will be compressed to form a neutron star which can be seen as the Pulsar (discuss in a future post) or an X-ray burster. In the case of most massive stars, the result is a black-hole that is larger that 4 M. In a neutron star , the material is present in the state called neutron-degenerate matter (discuss further) and also an complex form of degenerate material, QCD matter (discuss later) which is possibly inside the center.

The outer layers that are blown off of dying stars contain heavy elements that can be reused in the formation and growth of new stars. These heavy elements facilitate the creation of planets with rocky surfaces. Supernovae's outflow and the stellar winds of huge stars play a significant role in shaping the interstellar medium.

You can clearly see how complex the stars are, but you are aware of how awesome they are. But you aren't aware of something that I'll discuss later, like the QCD of pulsars and some other details I didn't discuss them at the in the same way because it will only add to the complexity of what already seem complicated, so it's the right time to learn about them too.

Radiation pressure refers to the mechanical pressure that is applied to any surface because of the movement of the object and electromagnetic field. This is the case for electromagnetic radiation or light of any wavelength that can be absorbed or reflected or else released (e.g. Black-body radiation) by matter at any scale (from microscopic things to particles of dust to molecules of gas). The force associated with it is called radiation force or in some cases, simply"light force.

White dwarfs, sometimes known as degenerate dwarf, can be described as a star-like core comprised of mostly electron-degenerate matter. White dwarfs are very dense, its mass is comparable to the mass of the Sun and its size is similar to the size of Earth. The white dwarf's dim luminosity is due to the release of thermal energy stored in the system; there is no fusion taking place in the case of a white dwarf.

Pulsars are magnetized , rotating compact star (usually neutron stars, but as well white dwarfs) which emits waves of electromagnetic radiation out through its poles magnetic. The radiation can be seen only when the beam of radiation is directed towards Earth (as similar to lighthouses, which is visible only when its light is directed towards you) and is the reason for the pulsed look of emission.

Quark (fundamental particle that is the basis of neutrons and protons are produced) matter, also known as QCD matter (quantum chromodynamic (it is a bit of a difficult subject, but I'll do my best to make it clear in the following the chapter)) can refer to one of a range of hypothetical matter phases that have degrees of freedom, including quarks and gluons one of which of quark-gluon plasma. There are a number of conferences planned for 2019 and 2020 as well as 2021 will be devoted to this particular topic.

The term "black dwarf" refers to a stellar remnant which is specifically white dwarfs that have been sufficiently cooled that it is no longer emitting significant light or heat. The time it takes for white dwarfs to reach the state of being a black dwarf is believed to be more than the timeframe of our universe (13.77 billion years) No black dwarfs are anticipated to be present within the universe up to this moment as their temperature is the same for the coldest white dwarfs is a possible observational limit to the age of the universe.

Quark stars are kind of hypothetical super-compact, exotic star in which extremely high temperatures and pressure has compelled nuclei to create quark matter. It is a constant state of matter made up of quarks that are free.

Binary stars are system of stars composed of two stars which orbit around their barycenters. Systems with multiple stars is known as multi-star systems. These systems, particularly when they are further away, typically appear to the naked observer as a single point of light. However, they can be seen as multiple through other means.

When stars are formed within the current Milky Way galaxy they are comprised of around 71 percent hydrogen, and 27% Helium according to mass. They also contain a small proportion that is heavier components. Typically , the proportion in heavy elements can be determined using the amount of iron in the stellar atmosphere since iron is a typical element and absorption lines are fairly easy to gauge. The proportion of heavier elements can indicate the possibility that the star is part of an planetary system.

Nucleosynthesis is the process by which one creates new atomic nuclei by combining nucleons that already exist (protons as well as neutrons) and nuclei. According to current theories the nuclei that first formed were created just a few minutes after the Big Bang, through nuclear reactions, a process known as Big Bang nucleosynthesis. Big Bang nucleosynthesis is the creation of nuclei that are not the ones found in the lightest of isotopes of hydrogen (hydrogen-1 1H, which has one proton in its nucleus) in the initial phases of the Universe. Primordial nucleosynthesis is believed to have occurred by the majority of cosmologists that it took place during the period of approximately 10 to 20 minutes following the Big Bang, and is believed to be the cause of the creation of the bulk of the universe's helium , as the isotope of helium-4 (4He) together with tiny amounts of hydrogen-isotope, deuterium (2H or D) as well as the Helium isotope (3He) and the very tiny quantity of the lithium isotope (7Li). Alongside these stable nuclei there were two radioactive, unstable isotopes were also created which were tritium, the heavy hydrogen isotope ($3H$ or tritium ($3H$ or) and the beryllium isotope called beryllium-7 ($7Be$) However, these unstable isotopes were later degraded to $3He$ or $7Li$ and $7Li$, respectively, as mentioned above.

Stellar nucleosynthesis is the process of creating (nucleosynthesis) of elements that are chemical via nuclear fusion reactions inside stars. The process has been occurring since the beginning of the creation of lithium, hydrogen and helium in the Big Bang.

I want to tell you one more thing related to limits like Chandrasekhar limit there is one more useful limit which is known as Tolman-Oppenheimer-Volkoff limit (TOV limit) it is an upper bound to the mass of cold, nonrotating neutron stars, analogous to the Chandrasekhar limit for white dwarf stars you don't have to remember all the things you have to just understand as much as you can because you are not going anywhere nor even book you can come back and have a look I am saying all this because it will help you to understand more complex topics in further chapters.

A astronomical unit (symbol au, or the letter AU) is a measure of length, which is roughly the distance between Earth towards the Sun and equivalent to around 150 million kilometers (93 million miles) or approximately 8 light minutes.

Chapter 4: Black Hole

The most famous celestial body that the majority of all over the world are aware of even if they weren't aware about it. They've have heard of it many times. The black hole is a an everyday name for Astronomers, physicists and physics fanatic. If anyone doesn't know about it, they'll search it, and they will discover a an intricate definition of black holes However, you're studying this book to comprehend complicated concepts in the similar way to how I learned it.

Then, what is a black hole? In simple terms, it's simply a space-time zone in space, where gravity has become so intense that no thing (no particle, or just light) is able to escape. Understanding the black hole is complicated, but if we attempt to get to know about the subject from the beginning (formation) in the first place, grasping the subject is simpler for us.

Gravitational collapse happens when the pressure inside an object isn't enough to counter gravity's pull on the object. For stars, this is usually due to a star having not enough "fuel" remaining to keep its temperature by undergoing nucleosynthesis in the stellar atmosphere, or the star that should be stable is flooded with extra matter in a manner that doesn't raise the temperature of its core. In any case, the star's temperature isn't sufficient to stop the collapse of the star under the weight of its own. The collapse could be prevented by the degeneracy force of the stars constituents and allowing the condensation of matter into a more dense state. The result is one the many kinds of compact stars. What type is formed depends on what the size of remnants from the original star in the event that the outer layers are removed (for instance in the case of a Type II supernova). Its mass, also known as the one that collapses and survives the explosion, could be significantly less than the mass of the star that was originally. Remnants with a mass of more than 5M are generated by stars that had a mass of over 20 million before their collapse.

If the mass of the remnant exceeds about 3-4 M (the Tolman-Oppenheimer-Volkoff limit), either because the original star was very heavy or because the remnant collected additional mass through accretion of matter, even the degeneracy pressure of neutrons is insufficient to stop the collapse. The mechanism that is currently in use (except the possibility of quark degeneracy pressure - see quark star) is strong enough to stop explosion and the object will eventually shrink to create an black hole.

The gravitational collapse of massive stars is thought to be the cause of the creation of black holes that are massive in the stellar universe. Star development at the time of the universe's beginning might have produced massive stars that after their collapse could result in black holes with a mass of up to 103 million. These black holes could have been the seeds of the massive black holes that are found in the center of many galaxies. There is also the possibility the possibility that huge black holes that have average masses of around 105 million may have been formed by an explosion of the gas clouds in the early universe. Certain possibilities for these objects have been identified through observations of the early universe.

The majority of the energy released by gravitational collapse is released quickly, the outside observer is not able to see the final result of this process. Although the process takes only a short period of time, if you look at the point of reference of falling matter, an observer from a distance could see the material falling slow and slow to a stop at the event horizon due to the gravitational dilation. The collapsing light will take longer and longer for it to get into the eyes of an observer as the light released prior to the moment when the event horizon is formed being delayed by for an indefinite amount of time. So, the observer outside is not able to see the creation of the event's horizon. instead the collapsing material appears to get dimmer and more shifting in red, before eventually disappearing.

Gravitational collapse isn't the only method that can produce black holes. In theory black holes could form by high-energy collisions with enough density. In 2002, there was no evidence of incidents of this kind have been identified either directly or indirectly due to due to a lack of mass balance of particle accelerator tests. This implies that there should exist a lower limit on Black holes' mass. Theoretically, the boundary is predicted to be close to approximately the Planck mass, at which point quantum effects are predicted to defy the predictions from general relativity. This would make the creation of black holes beyond the reach of any high-energy event that occurs near or on the Earth. However, some advancements in quantum gravity indicate the black hole's mass may be lower than that as certain braneworld (Brane Cosmology refers to a variety of theories of particle physics as well as the cosmology that is related to superstring theory, string theory theory, and M-theory.) For instance, some scenarios place the boundary as low to 1 TeV/c2. It is possible that micro black holes could be formed in high-energy collisions which occur when cosmic rays strike the Earth's atmosphere, or within CERN's Large Hadron Collider in CERN. These theories are highly speculation, and the development of black holes as a result of these kinds of processes is considered unprobable by many experts. Even even if micro black holes might be created the expectation is that they will disappear within 10-25 seconds and would not pose a danger to Earth.

After a black hole has been created, it will continue to expand by absorbing more material. A black hole can continuously absorb interstellar dust and gas from the surrounding space. This process of growth is a possible mechanism by which supermassive black hole may be formed, though the process of forming supermassive black holes remains an unexplored area of study. Similar mechanisms have been proposed for the creation of black holes with intermediate mass in the globular clusters. Black holes also have the ability to merge with other objects, such as stars, or even with different black holes. This may be important, particularly during the initial development of supermassive black holes which may have arisen by the aggregation of several smaller objects. This process is also suggested as the source of intermediate-mass black holes.

This is a debate about, and if there's no any mention of (I would like to honor him by adding legend as the suffix of his name) the legend Professor. Stephen Hawking then either the discussion is not complete or you did not have enough knowledge about the black hole.

The most well-known discoveries made by sir Stephen Hawking is the hawking radiation. We will talk more about it in the future this time. We learned a bit about it, so the radiation is black-body and is believed to release by black-holes because of quantum phenomena near the event horizon of the black hole.

As I have already said, when something was born in universe , it must die it is also applicable in black holes too, however it isn't yet fully proven dying black hole is commonly known as evaporation from black hole let's look at how they work.

in 1974 Hawking claimed that the black hole aren't completely black, but they emit small quantities of thermal radiation at temperatures of $c3/(8\ P\ G\ M\ kB)$ (ignore formulas if don't understand the concept this is just for those who are smarter than we) The effect is now called Hawking radiation. Through applying the quantum field theory on a static black-hole background, Hawking discovered that a black hole would emit particles that have an ideal spectrum of black bodies. Since Hawking's work, many other researchers have verified the conclusion using different methods. If Hawking's theory about black hole radiation is accurate the black holes tend to shrink and then evaporate when they lose mass due to the emission of photons as well as other particles. The temperature of this spectrum (Hawking temperature) is proportional to the gravity of the surface of the black hole that is, in the case of the Schwarzschild black hole is ininverse proportion to the mass. Therefore, larger black holes produce less heat than smaller black holes.

An astronomical black hole that is 1 M has 1 M has a Hawking temperature of the equivalent of 62 nanokelvins. This is much lower that the 2.7 Kelvins of radio waves from the background cosmic microwave. Black holes with greater mass have more mass in the background of cosmic microwaves than they emit via Hawking radiation and therefore they will expand rather than shrink. For an Hawking temperature higher that 2.7 Kelvin (and be capable of melting) the black hole must have to have a mass lower than that of the size of the Moon. A black hole of this kind has an area of less than 10 tenths of millimeter.

When a black-hole is small, its radiation effects could be extremely strong. A black hole of the size of a car could be about 10-24 meters and would take about a nanosecond to completely evaporate, in this time, it will possess a luminosity greater 200 times the luminosity is the Sun. The black holes of lower mass are predicted to melt faster than they did as an example that a black hole with mass 1 TeV/c2 could require less than 10-88 seconds to completely evaporate. For a black hole quantum gravity influences are predicted to play a major impact and may even allow a small black hole stable. However, the current quantum gravity developments don't suggest this is the scenario.

The Hawking radiation from the astrophysical black holes is believed to be fragile, and therefore very difficult to observe from Earth. One possibility however is the flash of Gamma rays that are released during the final phase of extincting of black holes from the primordial. Research into such flashes has failed and put strict restrictions on the possibilities of existence of primordial black holes. NASA's Fermi Gamma-ray Space Telescope launched in 2008 will continue to hunt to find these flashes.

When black holes melt away through Hawking radiation A supermassive solar black hole is likely to disperse (beginning at the point that temperatures of the cosmic microwave background falls to the level in the case of the black holes) over the course of 1064 years. A supermassive black hole that has an estimated mass of 1011 million will be gone in about 2x10100 years. The massive black holes of the universe are expected to continue to expand to 1014 M or more during the demise of superclusters of galaxies. These would also disappear in a period of time of up to 10106 years.

Humanity has recently accomplished one of the greatest feats in human history. It was that they captured the image of the black holes. They had accomplished this by organizing the team that would take on the challengeof creating the network of telescopes that is known by the name of Event Horizon Telescope which is also known as the EHT. The team set off to create the image of a black hole, by developing the technique that permits the capture of distant objects, referred to as Very Long Baseline Interferometry, or VLBI.

The aperture of a huge virtual telescope, such as that of the Event Horizon Telescope is nearly as big than the space between two telescopes that are the furthest apart in the EHT two stations are located at the South Pole and in Spain which creates an aperture almost the same size as the size of Earth. Each telescope within the array is focused on the object which is in this case, that's the black hole. It then collects information from its position on Earth and provides a small portion of the EHT's view. The more telescopes within this array are separated, the higher the resolution of the image.

In 2017 it was the time that in 2017, the EHT was a partnership of eight locations around the globe. More have been added since. In order for the group to start recording data they needed find a date when the weather would allow telescope viewing at all locations. In the case of M87*, they searched to find a good day in April 2017. out of 10 dates they selected to observe, five days had clear weather at each of the eight sites.

While NASA observations could not precisely trace the historical image, astronomers relied on information obtained from Chandra as well as NuSTAR satellites to determine the X-ray luminosity from M87*'s jet. Researchers used this data to evaluate their theories of the jet and disk surrounding the black hole to EHT observations. Additional insights could emerge as scientists continue to pore through the information.

Kudos for Katie Bouman who led the development of a new algorithm that was able to create the first-ever picture of the black hole

And here's the thing.

They are working harder to make images clearer and more effective. Here it's what they have made of the donut shape black hole. It's now more noticeable.

The most well-known and longest-running debates on the Black Hole was Black Hole Information Paradox ....

Some notorious physicists even made bets on the outcome after riddle will solved like- The Thorne-Hawking-Preskill bet was a public bet on the outcome of the black hole information paradox made in 1997 by physics theorists Kip Thorne and Stephen Hawking on the one side, and John Preskill on the other, according to the document they signed 6 February 1997

Since a black hole is characterized by only a couple of internal parameters, the bulk of the data about the material that was involved in the formation of the black hole goes unnoticed. No matter what kind of matter that forms the dark hole, it seems that only information regarding the mass as well as the charge and angular momentum remain. For as long as black holes are believed to be permanent, this information disappearance isn't a problem because the information can be interpreted as being within the black hole and is not accessible from outside, but visible on the horizon of the event in keeping with the holographic concept. However, black holes gradually fade away through emission of Hawking radiation. The radiation doesn't appear to contain any additional information on what caused the formation of the black hole. This means that the information is believed to disappear for good.

The issue of whether information really disappears inside black hole (the Black Hole Information Paradox) has split the theoretical physics community into quantum mechanics. The loss of information is a result of violation of a property known as unitarity. It's been argued that the loss of unitarity could also mean a an infraction to conservation of energy,[206however this has been disputed.Over recent years , evidence is mounting that information and unitarity are both preserved in a complete quantum gravitational analysis of the issue.

One approach to resolving the black hole paradox is called the black hole complementary theory. In 2012"firewall paradox" was introduced in 2012 "firewall paradox" was first introduced to show that black hole complementarity is not able to resolve the problem of information. Based on quantum field theories for the curved spacetime, one emitting of Hawking radiation is caused by two mutually connected particles. The particle that escapes is released as the quantum of Hawking radiation. The falling particle is swallowed up by the black hole. Imagine a black hole was existed for a finite period in the past. It will completely disappear within a finite period of time in the near future. It will then emit a limited amount of the information that is encoded in the Hawking radiation. Based on research conducted by physicists such as Don Page and Leonard Susskind There will come an era when the outgoing particle will be bound by all of the Hawking radiation that the black hole previously released. This could create an unsolved problem: a principle known as "monogamy of Entanglement" is that, like every quantum systems, an particle that is outgoing is not able to be completely interspersed with two other systems simultaneously however, the particle that is outgoing appears to be interspersed with the falling particle and independently with the past Hawking radiation. To resolve this issue, physicists might eventually be required to abandon one of the three well-tested theories that Einstein's equivalence principle is one of them or unitarity, or even the local quantum field theory. One solution that is in violation of the equivalence principle is that"firewalls "firewall" blocks particles from entering on an event's edge. The question is, what of those assumptions ought to be discarded is a subject of discussion?

Hawking suggested that information disappears in black holes and is not preserved by Hawking radiation. Susskind was not convinced, and argued that Hawking's theories were in violation of one of the most fundamental laws of science in the universe, which is the preservation of information. According to Susskind writes in his book The Black Hole War was an "genuine scientific debate" among scientists who favor the fundamentals of relativity and those who favored quantum mechanics. The debate resulted in the concept of holographics, first suggested by Gerard 'tHooft and further refined by Susskind who suggested that the information in fact is retained, stored at the edges of the system.

If you are interested in the fight of great physicists over the black holes, I'd like to suggest you to read the Sir Leonard Susskind book THE BLACK HOLLE WAR. I also recommend that you watch the universe functions in season 1, episode 2. can help you understand black holes more clearly.

In the 10th grade, I wasn't a one who was a physics enthusiast, and I disliked it due to its insolvable numerical. At the moment of my annual test I was looking for something on Google and when I came across the article about 10 fascinating facts about black holes. I clicked on it, and after reading the article, I thought, what's that? It is possible to find something similar in the universe, and I had no idea about it until I was introduced to the big bang (discuss in the next chapter) and I felt like I'm obsessed with physics. I enrolled in the science stream in the class of 11 and afterwards I was thinking that I need to find out more about astronomy as well as all other kinds of physics that exist. It is my goal to publish an ebook on the most fascinating subject of all time (physics) since

I'm sure there are many people who want to know more about it, without having to memorize formulas, formulas, or derivations. mathematical.

That's the reason I have a an affinity for big bang and black holes as they are the primary reason for me to be fascinated by Physics. We'll return to the subject of understanding black hole. We need to look at the structure of the black hole beginning with the event horizon, then singularity, photosphere and ergosphere.

Event Horizon -- The most distinctive feature of black holes can be seen in the form of an event-horizon, a limit in spacetime that light and matter are able to pass only towards the massive black hole. Nothing is able to escape, not even light is able to escape an event horizon. The term "event horizon" refers to as such because , if an event occurs within its area, the information generated by the event will not be transmitted to an observer outside which makes it difficult to establish whether an incident occurred.

According to general relativity an object of mass alters spacetime in a manner that the paths of particles are bent towards the mass. On the horizon of the event that is the black hole this deformation gets so powerful that there is no path that diverge towards the dark hole.

For a distant observer clocks close to an amorphous black hole will appear to be running slower than clocks further to the black holes. Because of this effect, known as gravitational dilation, a object that falls into an black hole appears slow down as it nears the horizon of the event, taking an infinity of amount of time to reach it. However everything else on the object slow down from the perspective from a fixed observer which causes any light that is emitted through the spacecraft to look brighter and dimmer, a phenomenon referred to as gravitational redshift. In the end, the object is gone, and it is no longer observed. The process usually happens quickly, with objects disappearing completely from view in less than an instant.

However, those who are indestructible and fall into a black hole will not experience any of these consequences when they reach the event horizon. Based on their own clocks which appear to be normal and then cross the event horizon in some time and do not notice any specific behavior. In the classical general relativity it is not possible to pinpoint the whereabouts that the event's horizon is using local observations, because of Einstein's equivalence theory.

Topology for the event horizon the black hole in equilibrium is always the shape of a spherical. In those that are not rotating (static) black holes,, the topology of their event horizons is the same, but for rotating black holes, the event Horizon is an oblate.

Singularity -- In the heart of a black hole according to general relativity, is the gravitational singularity, which is an area where the curvature of spacetime becomes infinite. In the case of a black hole that is not rotating it takes the form of a single point , and in the case of a rotating black hole it is dispersed into a ring singularity that is situated within the plane of rotation. In both instances the singular region is zero volume. It is also evident that the singular region is home to all the weight of the black hole solution. The singular region may therefore be imagined as having an infinite density.

The observers who fall into the Schwarzschild black hole (i.e. non-rotating and uncharged) can't avoid being carried into the singularity after they reach the threshold of the event. They are able to prolong their experience by speeding up to slow their descent but only to a point. Once they reach the point of no return and are in a state of collapse until they reach an infinite density. their weight is added to the mass in the dark hole. Prior to that they'll have been disintegrated by the expanding force of the tidal in the process known as the spaghetti effect or "noodle effects".

In the event of an charge-driven (Reissner-Nordstrom) as well as a moving (Kerr) black hole It is possible to circumvent the singularity. The extension of these solutions as far as is possible, we can see the possibility of exiting from the black hole and entering a new spacetime , with it acting like an"wormhole. The possibility of travelling to another universe is however, purely theoretical, as any change in the universe would eliminate the possibility. It is also possible to create closed time-like curves (returning to one's own time) in the vicinity of Kerr singularity, which can lead to problems with causality , such as that of the grandfather paradox. It is likely for these bizarre phenomena would be sustained in the correct quantum treatment of charged and rotating black holes.

Photon sphere The photon globe is a spherical border that has no thickness. photons that travel along tangents to the sphere will be caught inside a circular orbit around the dark hole. In the case of black holes that do not rotate the photon sphere has the radius of 1.5 times that of the Schwarzschild radius. Their orbits are dynamically unstable. Any tiny perturbation, like particles of falling matter could cause an unstable state that would expand in time, setting the photon onto an outward direction, causing it to flee from its black hole. Or, it would set taking it on an inward spiral , where it will eventually cross over the boundary of the event horizon.

Even though light may escape the photon sphere and enter the photon sphere, any light that crosses the photon sphere along an inbound path is taken from the black hole. So any light that gets to an observer who is not inside the photon sphere should be produced by objects that lie between the photon sphere as well as the event horizon. In the case of an Kerr black hole, the size of the photon's sphere is contingent of the spin parameters, and on the specifics of the photon's orbit that can be either retrograde (the photon's orbit rotates in the same way as the spin of the black hole) and retrograde.

Ergoshpere The black holes are enclosed by a spacetime area that is unable to remain still, referred to as the Ergosphere. This is due to frame-dragging. General relativity states that any object that is rotating will be inclined to "drag" across the spacetime that surrounds it. Anything that is close to the object that is moving will have a tendency to begin moving towards the direction of the rotation. For a black hole that is rotating the phenomenon is so powerful near the event horizon an object will need to be moving more quickly than light moving in the opposite direction in order to remain stationary.

The ergosphere in a black hole is an area bound by the event horizon of the black holes and the ergosurface. The ergosurface is located at the poles, however it is located at a greater distance from the Equator.

The radiation and objects that escape in the normal way out of the ergosphere. By this process, known as the Penrose procedure, the objects may escape from the ergosphere using more energy than they came in with. The additional energy comes from the energy that rotates the black hole. The black hole's rotation is slowed down. The variation on the Penrose process occurs in the presence of magnetic fields, called the Blandford-Znajek procedure is thought to be a plausible explanation to explain the huge luminosity as well as relativistic jets of quasars as well as other galactic nuclei that are active.

Chapter 5: Big Bang

The big bang theory is the basis for the universe that is visible. This model explains how the universe grew after a beginning state with high density temperatures and provides a thorough explanation of a wide range of observed phenomena. These include the numerous light elements, radiation of the cosmic microwave background (CMB) radiation and the large-scale structure.

In the Big Bang theory, the expansion of the visible universe was initiated by the explosion of one particle at a certain moment in time.

The idea was first presented in the form of a scientific paper in 1931 in a paper written published by Georges Lemaitre, a Belgian scientist as well as a Catholic priest. The idea, which is believed by the majority of astronomers in the present is an unorthodox departure from the traditional scientific thinking at the time of the 1930s. A lot of astronomers in the 1930s were skeptical of the notion that the universe was expanding. The idea that the universe of galaxies started with a bang seemed absurd.

The theory is based on two main theories: the universality physical laws as well as the cosmic principle. Physical laws are universal and one the foundational principles of Relativity theory. The cosmological principle states that on a large scale, all the world's phenomena are homogenous, and isotropic, appearing identical in all directions regardless of where it is located.

Expanding the Universe was discovered by observations of astronomy and is an integral component in the Big Bang theory. Mathematically, general relativity defines spacetime using a metric which defines the distances that exist between adjacent points. Points, that may be galaxies, stars or any other object are defined by a coordinate chart or "grid" which is laid out across all spacetime. The cosmological principle implies that the metric should be homogeneous and isotropic on large scales, which uniquely singles out the Friedmann-Lemaitre-Robertson-Walker (FLRW) metric. This metric includes an element called a scale factor. It defines how the dimension of the universe alters as time passes. This allows a logical choice for a system of coordinates be created, referred to as comoving coordinates. In this system of coordinates the grid expands with the universe. objects that move due to the expanding universe remain fixed in the grid. Although their distance to coordinate (comoving distance) remains constant however, physically distances between these moving points increases proportionally to the magnitude factor of the universe.

The Big Bang is not an explosion of matter expanding outwards to fill the universe. Space expands in all directions and also increases in physical distance between the comoving points. This means that it is not the case that the Big Bang is not an explosion that occurs in space, but instead the expanse of space. Since the FLRW measurement relies on a uniform distribution energy and mass and energy, it is applicable to our universe on only larger scales. Local quantities of matter, such as our galaxy are not required to expand at the same rate that the rest of the Universe.

One of the most important features that is a key feature of Big Bang spacetime is the existence of particle Horizons. Since the universe is an age that is finite and light moves at a certain speed, it is possible that there are instances in the past that light has not had enough time for it to get here. This puts a limit or a past horizon for the objects with the greatest distance that are visible. However, as the universe is expanding and objects that are further away are receding more rapidly our light today will be unable to "catch with" to distant objects. This is referred to as the term "future horizon," which restricts the kinds of possibilities of future events that we are able to influence. The existence of either kind of horizon is contingent upon the specifics in the FLRW model that defines our universe.

Our knowledge of the universe going back to the beginning of time suggests there is an horizon that is past, but in reality, our understanding is limited by the opacity of the universe in the beginning of time. Therefore, our view is not able to extend further backwards in time even though the horizon is receding in space. As the Universe continues to expand there will be a new horizon, too.

The extrapolation of the expansion of the universe backwards through time by using general relativity results in an infinite temperature and density at a time that is finite within the last. This strange behavior, also known by the name of gravitational singularity (is an area in spacetime in which the gravitational field as well as density of a celestial body are predicted to increase to infinite levels by general relativity, in a manner which is independent of the system of coordinates.) This indicates that general relativity isn't the best explanation of the physics laws in this particular regime. The models built on general relativity alone cannot be extrapolated to the singularity before the conclusion of the Planck period (describe the evolution and history of the universe according to the big bang Cosmology).

The initial stages that formed the Big Bang are subject to lots of speculation because the astronomical evidence about them is not available. According to the most commonly used models it was believed that the universe was full uniformly and isotropically, with a massive energy density, as well as enormous pressures and temperatures, and was expanding rapidly and cooling. The time between 0 and 10-43 seconds of expansion, known as the Planck period was a time where the four major forces, the electromagnetic force strong nuclear force and the nuclear force that is weak along with the gravitational forces were joined together as one. In this phase the typical scale size of our universe is the Planck length of 1.6×10^{-35} millimeters, and was a temperature of around 1032 degrees Celsius. The very notion of a particle collapses under these conditions. A complete understanding of this time period is needed for the formulation of a quantum theory gravity. The Planck period was followed with the grand unification period that began at 10-43 second, when gravity separated itself from other forces of the universe as its temperature decreased.

Around 1037 second into expansion, the phase change resulted in a cosmic explosion, in which the universe expanded exponentially, free of the invariance of light speed and temperatures decreased by 100,000. The microscopic quantum fluctuations that took place due to Heisenberg's uncertainty principle were amplified to create the seeds that later become the larger-scale world structure. In a period of 10-36 seconds the electroweak epoch is with the strong nuclear force is separated from the other forces, with the electromagnetic force as well as the weak nuclear force remaining united.

The rate of inflation slowed down about the 1033 to 1032 second mark, the volume of the universe having grown by at least 1078. Reheating continued until the universe had reached the temperature required for creation of the quark-gluon plasma as along with the different elementary particles. Temperatures were so high that the random motions of particles were at relativistic speeds, and particle-antiparticle pairs of all kinds were being continuously created and destroyed in collisions. Somewhere, an unknown process known as the baryogenesis (is an atomic process thought to have occurred in the beginning of the universe, resulting in baryonic asymmetry i.e. the inequities between material (baryons) as well as matter (antibaryons) within the universe observed.) The baryons' conservation was violated. number, resulting in the existence of a tiny excess between quarks and leptons versus antiquarks as well as antileptons, of the order of one-third of a million. This resulted in the dominant dominance that matter is superior to antimatter within our present universe.

The universe continued to shrink in density and drop in temperature, and the normal energy of particles was declining. Symmetry-breaking phase transitions bring in place the fundamental forces of Physics as well as the elementary particle parameters into their present forms and magnetic force as well as the weak nuclear force splitting around 10-12 seconds. After 10-11 seconds, the picture is more definite, as the energy of particles decreases to levels that are achievable in particle accelerators. In about 10-6 seconds, gluons and quarks (is an elemental particle that functions as an exchange particle (or gauge boson) for the strong force between quarks.) They are combined to create baryons (is a kind of subatomic particle that includes an odd number of quarks with valence (at at least three).) like neutrons and protons. The slight surplus of antiquarks and quarks resulted in a slight excess in baryons over antibaryons. The temperature was now no longer high enough to create new proton-antiproton pairs (similarly for neutrons-antineutrons), so a mass annihilation immediately followed, leaving just one in 108 of the original matter particles and none of their antiparticles. The same thing happened in about a second for positrons and electrons. After the annihilation all the electrons, protons and neutrons weren't moving relativistically anymore as the density of energy in the Universe was controlled by photons (with some contribution from neutrinos).

Within a couple of minutes of the expansion, at a time when the temperature was around 1 billion kelvin as well as the mass density in space was similar to the density currently found in Earth's atmosphere, neutrons merged with protons to create the universe's deuterium and

Helium nuclei through a process known as Big Bang nucleosynthesis (BBN). The majority of protons were not incorporated into hydrogen nuclei.

As the universe warmed and the energy density of matter began to dominate the radiation from photons. After 379,000 years, the nuclei and electrons combined to form Atoms (mostly hydrogen) that could emit radiation. The relic radiation, which was able to travel across space virtually unimpeded, is referred to as cosmic microwave background.

Evidence from independent lines of inquiry that come from Type Ia supernovae and the CMB indicate that the universe of today is controlled by a mysterious type of energy referred to as dark energy. This form of energy appears to permeate all the space. The evidence suggests that 70% of the energy density of the universe today is in this type of form. In the early days of the universe, when it was young, it was probably packed with dark energy however, with smaller space and everything close to each other, gravity dominated and was slowing the expansion. After a few billion years of expansion the decreasing density of matter compared the dark-energy density led to the universe's expansion to slow down and increase.

Dark energy, in its simplest formula is the cosmological constant that is part of Einstein fields equations for general relativity however its mechanism and composition are not known. More broadly, the particulars the equation and its relationship to the Standard Model of particle physics remain to be studied by observation and theoretically.

The whole cosmic process after the inflationary epoch could be precisely described and modeled using the LCDM model of cosmology which is based on the distinct structures of quantum mechanics as well as general relativity. There are no testable models that could be able to describe the event prior to about 10-15 seconds. Understanding the earliest period in the evolution of space and time is one of the most difficult questions in the field of the field of physics.

Measurements of the redshift-magnitude ratio for supernovae of type la show that the universe's expansion has been increasing as the universe was around half its current age. To explain the speed, the theory of general relativity suggests that a large portion of the energy that is circulating in the universe is composed of a substance with a significant negative pressure, which is known as "dark energetics".

Dark energy, although speculation, can solve a variety of problems. The measurements of the background of cosmic microwaves indicate it is close to being flat spatially which means that according to general relativity , the universe must be nearly what is considered to be the crucial density in energy/mass. However, its mass-density is measured by the gravitational clustering of it, and it is estimated to be just 30 percent that of its critical density. The theories suggest that dark energy doesn't cluster in the normal way, it's the most plausible theory to explain this "missing" energetic density. Dark energy can also help clarify two geometric measurements of the curvature in the universe. One is based on an analysis of the gravitational lens's frequency and the second using the signature pattern of the massive structure as cosmic ruler.

Negative pressure is thought to be the result of vacuum energy. However, the precise nature and origin of dark energy is one of the most intriguing mystery that surround the Big Bang. Results obtained by WMAP's WMAP Team in 2008 agree with a universe made up of 73 percentage of dark energy as well as 23 percent dark matter 4.6 percent regular matter, and less than 1percent neutrinos. According to theories that the energy density of matter decreases as the size of the universe, however the dark energy density stays the same (or close to it) when our universe grows. So, matter constituted more of the universe's total energy earlier than it does now however its contribution will decrease in the near future, as dark energy becomes increasingly predominant.

The dark energy portion that makes up the universe is explained employing a range of theories, such as Einstein's cosmological constant, as well as more complex forms of quintessence, or other gravity theories that have been modified. The cosmological constant is often called the "most problematic physics problem" arises from the apparent contradiction between the observed energy density of dark energy as well as the one predicted by Planck units.

In the 1970s and 1980s, a variety of observations proved that there was not enough material visible in our universe that could explain the apparent force of gravitational forces that exist between and within galaxies. It led to the hypothesis that as much as 90 percent of the matter that exists within the universe is black matter which doesn't emit light or interact with baryonic matter that is normal. Furthermore, the notion that all matter in the universe was primarily regular matter led to theories that are in stark contrast to observations. Particularly the current universe, it is much larger and has much less deuterium than could be explained without dark matter. Although dark matter has always been controversial, it can be determined by a variety of observations such as the anisotropies of the CMB Galaxy cluster velocity dispersions massive-scale structures distributions and gravitational lensing research and X-ray studies for galaxy clusters.

Direct evidence of dark matter is derived due to its gravitational effect on other elements, since there are no particles of dark matter that have ever been observed in labs. Numerous particle physics theories to be dark matter candidates have been suggested and several research projects to discover them directly are in progress.

There are also a number of outstanding problems with the popular cold dark matter model , which include the dwarf galaxies problem and the cuspy-halo problem. Alternative theories are being proposed which do not require an enormous amount of unknown particles, rather they alter the gravity laws created through Newton and Einstein however, none of them have been as effective as Einstein's cold dark matter theory in explaining the observables of all time.

The magnetic monopole issue was first raised in the latter part of 1970. Grand Unified theories (GUTs) predicted space-time topological anomalies that would result in magnetic monopoles. These objects could be produced effectively in the early hot universe, leading to the density of these objects being much higher than that observed in the absence of monopoles, since none have been discovered. This issue can be solved with cosmic inflation. This takes away all defects that are point-based from the visible universe similar to the way it pushes the geometry towards flatness.

One of the most common misconceptions regarding The Big Bang model is that it can fully explain the origin and evolution of spacetime. But, it is not. The Big Bang model does not explain how time, energy and space came into being however, rather, it describes the creation of the present universe from an ultra-dense , high-temperature state. It's misleading to depict how the Big Bang by comparing its dimensions to the size of ordinary objects. When the universe's size during the Big Bang is described, it's referring to the dimensions of the visible universe and not the whole universe.

Hubble's law says that galaxies beyond Hubble distance shrink faster than light speed. But, special relativity does not apply to motion that is beyond space. Hubble's law describes the velocity that results from expanding space, not through space.

Astronomers typically refer to the redshift of the cosmological universe as an Doppler shift, which could cause confusion. Although they are similar however, the cosmological shift isn't the same as the conventionally calculated Doppler redshift since the majority of elementary methods of calculating the Doppler redshift are not able to accommodate the expanding of space. An accurate calculation from the cosmic redshift demands the application of general relativity and while a method based on simple Doppler effect arguments produces similar outcomes for galaxies near to each other interpret the redshift of galaxies that are more distant as result of the most straightforward Doppler redshift theories can lead to confusion.

The flatness issue (also called "the problem of oldness") can be described as an observational issue that is associated with an FLRW. The universe can be characterized by positive, negative or even zero curvature, based on its energy density. Curvature is negative when its density is lower than its critical density; positive when it is higher; and zero at the critical density. In this it is believed that space will be flat. The universe's observations are compatible with being flat.

The issue is that even a tiny deviation from the critical density is a gradual increase in density, however, the universe of today remains very flat. Since a normal timescale for a departure from flatness could be Planck time of 10-43 seconds and the reality that the universe has not reached a temperature death or the Big Crunch after billions of years of time requires an explanation. In particular, even at a relatively old age of a couple of minutes (the nucleosynthesis time) the cosmic density must be within one percent in

1014 of its crucial value or else it wouldn't be as it is today. Prior to the discovery about dark energy cosmologists thought of two possibilities that could be the future for the universe. In the event that the density of our universe was higher that the density of its critical then the universe would attain an absolute size, and then begin to shrink. It would get denser and hotter until it would end in a similar state to the one in the beginning--a Big Crunch.

In the alternative, if the density of the universe were at or less than the threshold density then it would expand slow but would not cease. Star formation would stop due to the depletion of interstellar gases in each galaxy. Stars would then disappear and leave neutron stars, white dwarfs as well as black holes. Collisions between them would result in the accumulation of mass into larger and bigger black holes. The temperature average of the universe will slowly asymptotically reach absolute zero, which is a Big Freeze. Furthermore, if protons were unstable the baryonic matter will disappear completely, leaving only radioactivity and black holes. In the end, black holes will disintegrate by producing Hawking radiation. The energy density of the universe will grow to the point that there was no form of energy that can be extracted. This is called heat death.

Modern observations of expanding at an accelerated rate suggest that more and more of the current visible universe will move beyond the horizon of our event and beyond our reach. The final outcome isn't yet known. It is unclear what the final result will be. LCDM version of our universe is based on dark energy in the form the cosmic constant. The theory implies that only gravitationally bound systems, like galaxies will remain connected, and are susceptible to death by heat in the course of the expansion and gets cooler. Different explanations for dark energy, known as ghost energy theories, claim that eventually the galaxy clusters nuclei, planets and atoms as well as the material itself could be pulled apart due to the constant expansion of the known as Big Rip.

It isn't understood what causes the universe to have higher amounts of antimatter than matter. It is believed that at the time when it was still young, and hot, it was at equilibrium statistically and had the same amount of baryons as antibaryons. However, evidence suggests that our universe and the most distant regions are made mostly of matter. The process of baryogenesis was proposed to explain the Asymmetry. To allow baryogenesis to take place it is necessary that there must be Sakharov prerequisites must be fulfilled. These stipulate that the number of baryons cannot be conserved and C-symmetry and CP-symmetry are not respected as well as that all the elements of our universe is diverging from the equilibrium of thermodynamics. All of these are found within the Standard Model, but the consequences aren't enough to account for the present baryon Asymmetry.

Future gravitational-wave observatories may be capable of detecting primordial gravitational waves, which are relics of the early universe for less than one minute in the aftermath of the Big Bang.

However, the Big Bang along with the inflation theory are not completely proven that it is the method by which the universe formed, but they are among the most reliable theories due to the fact that the astronomers who use this theory can solve problems despite the fact that it's making physicists ask additional questions that aren't solved by this theory yet perhaps the theories are correct but were not completely comprehended by humans.

There's a quote by Michio Kaku that is similar to this: First of all, Big bang was not that big. In addition, it didn't have a bang. Third Big bang theory can't provide any information about what was banged or when it banged. It only said there was a bang. Also, you can conclude that Big bang theory in same sense is completely a misnomer.

Chapter 6: Supernova

Supernova is an intense and bright stellar explosion. This astronomical event that is brief and fleeting occurs in the final evolution that a huge star undergoes when white dwarfs are triggered into nuclear fusion that is runaway. The object that started it all, known as the progenitor, eventually collapses into a neutron star, and black hole or becomes completely destroyed. The highest optical brightness of a supernova could be as high as the brightness of a galaxy before diminishing over months or weeks.

Theoretical studies suggest that the majority of supernovae are caused by one of two fundamental mechanism: the sudden re-ignition of nuclear fusion within the degenerate star, such as white dwarfs or the sudden collapse of the massive star's core. In the first type of instances, the temperature of the object increases enough to trigger nuclear fusion that is runaway which completely destroys the star. It could be due to the accumulation of materials in a binary partner through an accretion process, or an accretion of stellar material. In the case of a massive star the central center of the massive star could be prone to sudden collapse, which releases energy from gravitational potential that could result in an event known as a supernova. Although some supernovae observed are more complicated than these two simple theories The astrophysical mechanics have been recognized and accepted by the Astronomical community.

Supernovae may release many solar mass of material at as high as several percent of what light travels at. This causes an expansion of a shock wave that is absorbed by the interstellar medium, which carries an expanding gas shell and dust, which is referred to as the remnant of a supernova. Supernovae are an important source of elements found in the interstellar space from the oxygen to rubidium. In the course of expanding, shock waves from supernovae could trigger the creation in new stars.

Supernova remnants could be the main source cosmic radiation. Supernovae may cause gravitational waves. However, until now gravitational waves have been observed only through the mergers of neutron and black stars.

Astronomers categorize supernovae based on their light curves, as well as patterns of absorption from various chemical elements that are visible in their spectrum. If a spectrum from a supernova has hydrogen lines (known by the Balmer series within the visible part in the spectrum) it is classified as Type II; otherwise it is classified as Type I. In both kinds, there are subdivided in accordance with line patterns that originate from other elements, or the shape of the curvature (a graphic representation of the apparent magnitude of the supernova in relation to time).

Type I -- Type I supernovae can be classified by their spectrum, with type Ia displaying a powerful Ionised Silicon absorption line. Type I supernovae with no strong linear feature are classified types Ib and Ic Type Ib has solid neutral helium lines, and type Ic lagging the lines. The light curves are identical, though types Ia tend to be more brighter at the peak of their luminosity however, the light curve isn't crucial for the classification of types I supernovae.

There are a few types of Ia supernovae have unusual characteristics like different luminosity levels and broadened curves of light and they are often classified according to the first instance that displays similar characteristics. For instance this sub-luminous SN 2008ha is usually described as SN 2002cx-like or class IIa-2002cx.

A small percentage of the type Ic supernovae exhibit extremely broadened as well as blended emission lines that are believed to indicate extremely large expansion velocity for the exodus. They are classified as Ic-BL type or Icb.

Type II Supernovae of the type II could be subdivided on the basis of their spectrum. Although the majority of types II supernovae have large emission lines that show expansion speeds of hundreds of thousands of kilometers per second However, some, like SN 2005gl, exhibit narrow characteristics in their spectrum. These are known as type IIn. The "n" is a reference to "narrow'.

Certain supernovae, including SN 1987K, and SN 1993J, are believed to alter their types. They have hydrogen lines in the beginning of their lives, but over the course of months or weeks, get dominated by lines of Helium. "Type IIb" is the term used to describe "type IIb" is used to refer to the mix of features typically found in types II, and Ib.

Type II supernovae that have normal spectra, dominated by wide hydrogen lines that persist throughout the duration of the decline. Type II supernovae are classified based on the light curves they show. The most frequent type has an distinct "plateau" on the spectrum of light immediately after the peak brightness, and the brightness of the visible spectrum remains fairly steady for a few months prior to the decline begins. They are referred to as type II-P and refer to the plateau. The less common supernovae with type II-L but without an identifiable plateau. The "L" means "linear" but the light curve isn't necessarily straight.

Supernovae that are not able to belong to the usual classifications are referred to as peculiar or "pec".

Type III IV, VType III, IV, V - Fritz Zwicky defined additional supernovae types based on a few cases that did not exactly match the criteria for Type I or Type II supernovae. The SN 1961i supernova located in NGC 4303 is the first and the only member of the class of supernovas that was type III known for its large light curve's maximum as well as its wide hydrogen Balmer lines that took time to evolve within the spectrum. SN 1961f found in NGC 3003 is the prototypical and the sole member of the class type IV which had the same light curve as an II-P type supernova featuring hydrogen absorption lines and low emission from hydrogen. The class of type V was named for the SN 1961V observed in NGC 1058, a rare faint supernova , also known as a supernova impostor that has an incredibly slow increase in brightness, with a peak lasting several months and an uncommon emission spectrum. The similarity of SN61V with that of the Eta Carinae Great Outburst is also noted. Supernovae in M101 (1909) and M83 (1923 and 1957) were also considered to be possible types IV or type V supernovae.

All of them would be considered to be peculiar kind II supernovae (IIpec) and of which numerous more have been discovered, however it remains to be determined whether SN 1961V is actually a supernova due to an LBV explosion or an fake.

White dwarf stars can collect enough material from an astronomical companion to raise the temperature of its core to trigger carbon fusion in which case it will undergo the runaway nuclear fusion process, totally breaking the process. There are three ways that this detonation can be thought to occur: stable accumulation of material from a companion, collision of white dwarfs or accretion which causes the shell to ignite, which will then ignite the core. The primary mechanism through that the type la supernovae form remains elusive. However, despite this uncertainty regarding the way that they are produced, la supernovae are created they are of a the type la supernovae possess very consistent properties and serve as useful regular candles across intergalactic distances. Some calibrations are necessary to account for the gradual changes in properties , or the different frequencies of the supernovae that exhibit abnormal luminosity in high redshifts, as well as for slight variations in brightness that are detected by light curves or spectrum.

The collapse of the cores of massive stars could not produce visible supernova. The primary explanation is that there is a sufficient massive core with kinetic power that not enough to reverse the fall from the layers that surround it on the black hole. These kinds of events are hard to discern, but massive surveys have identified potential possibilities. The supergiant red N6946-BH1 of NGC 6946 experienced a small explosion in March 2009 but then faded away. A faint source of infrared light is visible at the location of the star.

The classification of supernovas is closely linked to the type of star at moment of collapse. The frequency of each type of supernova is dependent upon the metallicity, and consequently the age of the host galaxy.

Type Ia supernovae originate from white dwarfs in binary systems. They can be found across all types of galaxies. Supernovae that collapse into cores are discovered in galaxies in the process of an extremely recent or current star formation since they result from the short-lived massive stars. They are typically observed in types of Sc spirals, but they can also be found within the arm of spiral galaxies, as well as in galaxies that are irregular, particularly starburst galaxies.

Type II-L and Type Ib/c and perhaps the majority of types IIn, supernovae can only be believed to originate by stars that have near-solar metallicity levels , which result in a massive loss of mass from massive stars. This is why they are rarer in galaxies older and more distant.

In some supernovae that collapse into the core that fallback on the black hole causes relativistic jets that can produce an energetic and brief burst of gamma-rays. It can also transfer significant energy to the material that is ejected. This is one possible scenario for the production of high-luminosity supernovae. It is believed to be the main cause behind the type Ic hypernovae as well as long-duration gamma-ray bursts. If the relativistic jets are short and do not reach the outer layer of the star,, then an gamma-ray burst with low luminosity can be observed and the supernova could be sub-luminous.

When a supernova is discovered within an extremely dense cloud of material from the circumstellar region the result is an energy wave that will effectively convert a large portion of the energy kinetically generated to electromagnetic energy.

While it was normal, the resultant supernova will be bright and a longer duration because it is not dependent on the exponential decay of radioactive particles. This kind of event can result in the type IIn hypernovae.

While pair-instability supernovae are considered to be supernovae with core collapse that have spectrum and light curves that are similar to the type II-P supernova, the structure after the collapse of the core is more similar to that of the massive type Ia with an explosive fusion of carbon, oxygen and silicon. The energy release by the largest mass events is comparable to those of other supernovae with core collapse, however neutrino production is believed to be extremely low, which is why the energy kinetic and electromagnetic released is quite high. They have cores bigger than white dwarfs and the quantity of radioactive nickel and other heavy elements released from their cores may be several orders of magnitude greater and consequently, a high visible luminosity.

Supernovae are the primary source of elements within the interstellar medium , from oxygen to rubidium however, the actual abundance of the elements that are produced or seen in spectrum varies greatly based on the supernova type. Type Ia supernovae create primarily iron-peak and silicon elements including the iron and nickel. Supernovae that collapse into the core release less iron-peak elements than Ia supernovae. However, they produce more alpha elements that are light, such as neon and oxygen, and elements that weigh more than zinc. This is particularly the case when it comes to Supernovae that capture electrons. The majority of the material produced by supernovae of type II is helium and hydrogen. The heavy elements are formed through nuclear fusion for nuclei with a size of 34S or greater the rearrangement of silicon photodisintegration and quasiequilibrium during the silicon burning of nuclei between 56Ni and 36Ar; and the rapid absorption from neutrons (r-process) in the collapse of the supernova, for elements that are heavier than iron. The r-process results in extremely unstable nuclei that are high in neutrons, and are rapidly decaying to more stable types. In supernovae, r processes can be responsible for approximately half of the isotopes of elements other than iron, even though neutron star mergers could be the main source of many such elements.

In the present universe, the old asymptotic giant branches (AGB) star clusters are the main source of dust derived from oxides, s-process elements and carbon. But in the early universe, prior to when AGB stars were formed supernovae could have been the primary source of dust.

Supernova remnants may increase the speed of a significant portion of galactic cosmic rays that are primary However, evidence of cosmic rays has only been observed in a limited amount of the remnants. Gamma rays of pion-decay have been observed in Supernova debris IC 443 as well as W44. They are created when protons are rapidly accelerated from the SNR collision with interstellar matter.

Supernovae can be strong cosmic sources of gravitational waves However, none have been observed. The only gravitational waves to date have been observed are mergers of neutron and black hole stars, which are likely remnants of supernovae.

A near-Earth Supernova is one that is that is close enough to Earth to cause noticeable changes on the biosphere. Based on the nature and the energy associated with the explosion, the distance may be as close as 3000 light years away. In 1996, it was speculated that the remnants of previous supernovae could be visible on Earth by way of metallic isotopes found in the strata of rock. The enrichment of iron-60 was later discovered in deep-sea rocks from in the Pacific Ocean. In 2009, higher levels of nitrate ions were observed in Antarctic Ice, which was in conjunction with the 1054 and 1006 Supernovae. Gamma radiations from these supernovae might boost the levels of nitrogen oxides which were trapped within the glacier.

Type Ia supernovae can be believed to be among the most dangerous , if they are close enough to Earth. Because they originate from common, dim white dwarf stars that are found in binaries, it's possible that a supernova that could impact the Earth could occur without warning and in a system of stars which isn't well-studied. The closest candidate to be identified is IK Pegasi . Recent research suggests that the type II supernova will need to be more to eight Parsecs (26 lights-years) to completely destroy the Earth's ozone layer and there aren't any candidates with a distance of less than 500 light years.

For those who aren't sure what parsec is, what do you know? 1 parsec equals 3.25 Light years.

Scientists have discovered many things about the universe from studying supernovas. They employ the second kind of supernova (the one that involves white dwarfs) as a ruler to gauge distances within space.

They also have discovered they are also the factories of the universe. Stars create the chemical elements required to create everything in our universe. They are the stars that transform simple elements like hydrogen to heavier components. The heavier elements, like nitrogen and carbon are the necessary elements for living.

The only massive star can produce heavy elements such as silver, gold and the element uranium. If supernova explosions occur stars release both the stored as well as newly-created elements around space.

NASA scientists employ a variety of different telescopes to find and investigate supernovas. One of them can be the NuSTAR (Nuclear Spectrum Telescope) project, that makes use of an X-ray telescope to explore the universe. NuSTAR helps scientists study supernovas as well as young nebulas, to understand what happens before and following these amazing explosions.

Recent research has revealed that supernovas resonate like huge speakers, emitting an audible sound before they explode.

Researchers in 2008 observed an explosion of a supernova as it was in the process of exploding the first time. As she sat in front of her computer monitor the astronomer Alicia Soderberg expected to see the tiny, glowing smudge that was the supernova's month-old. However, what she and her colleague observed instead was an unusual intensely bright, five-minute flash of radiation.

After the discovery scientists became the very first to observe the star in the process of explosion. The supernova's name was SN 2008D. Additional research has confirmed that the supernova has unique characteristics.

"Our observations and models demonstrate this to be an unique event that should be understood more clearly by referring to an object located in the middle between normal supernovae and gamma-ray explosions," Paolo Mazzali, an Italian astrophysicist from Padova Observatory and Max-Planck Institute for

Astrophysics, Padova Observatory and Max-Planck Institute for Astrophysics has told Space.com in an interview with Space.com in 2008.

Chapter 7: Galaxy

Many thousands of solar systems similar to ours, with an underlying parent star that is gaseous and rocky planet that is removing from it. I'm not certain about the existence of extraterrestrial beings or even aliens, but I do believe that living beings exist in galaxies. They have star nurseries, where stars are born. There are also planets that don't revolve around stars. There are also numerous stars that don't have a solar system.

If we search for galaxy on google , we find that the definition of a galaxy is that it is one that is bound by gravitational force. It includes stars, stellar remnants interstellar dust, gas, and dark matter. of definition means that you know what galaxy is from my perspective and what is the definition of galaxy found on Google. It's all on to keep in mind.

Galaxies are classified based on their visual shape (is an approach utilized by astronomers to separate galaxies in groups based on their appearance.) as elliptical, spiral, or irregular. A lot of galaxies are believed contain supermassive black holes in their cores. The Milky Way's central black holes also called Sagittarius A*, has the mass of four million times larger than that of the Sun.

The current cosmological models of the beginning of the universe are built in the Big Bang theory. After about 300,000 years of this incident, hydrogen atoms and helium began to grow by Recombination. Most of the hydrogen was non-ionized (non-ionized) and easily attracted light. No stars had yet been formed. In consequence, this period is known as the "dark age". It was because of the fluctuation in density (or anisotropic disturbances) within this primordial matter that more complex structures began to form. The result was that masses of baryonic matter began to expand within cold dark matter halos. These primitive structures would eventually be transformed into the galaxies we observe now.

The precise process through the formation of the first galaxies remains a mystery in the field of astrophysics. Theories can be classified into two types which are top-down and bottom-up. In top-down correlations (such as the Eggen-Lynden-Bell-Sandage [ELS] model), protogalaxies form in a large-scale simultaneous collapse lasting about one hundred million years. In theories that are bottom-up (such as the Searle-Zinn model) smaller structures like large globular clusters are formed first and later, a few of these bodies merge to form a bigger galaxy. When protogalaxies started to grow and contract then the first halo star (called Population III stars) emerged within the clusters. They were made up mostly made of hydrogen, helium and other elements, and could have been more massive than 100 times Sun's mass. If that's the case, these massive stars would have rapidly consumed their fuel and would have become supernovae, which released massive elements from the space-time continuum. The first generation of stars re-ionized surrounding hydrogen and created large bubbles of space in which light was able to easily travel.

Within one billion years after the galaxy's birth, crucial structures begin to show up. Globular clusters (spherical collection of stars) and the central supermassive black hole, as well as galactic bulges of population stars that are metal-poor II stars are formed. The development of a massive black hole is believed to play an important role in controlling the growth of galaxies, by restricting the quantity of extra matter that can be that is added. At the beginning of this epoch galaxies go through a huge explosion of star creation. The process of star birth and death gradually increases the number in heavy elements eventually leading to the creation of planets.

The evolution of galaxies may be greatly affected by collisions and interactions. Galaxies merged in the beginning of the epoch and the vast majority of galaxies are unique in their morphology. Because of the distances between stars, the vast majority of star systems that are in collisional galaxies are unaffected. But, the gravitational stripping of interstellar gas as well as dust that compose the spiral arms results in the formation of a lengthy line of stars referred to as the tidal tails.

It is believed that the Milky Way galaxy and the nearby Andromeda Galaxy are moving toward each other at around 130 km/s. This means thatdependent on the lateral motions--they could meet in between five and 6 billion years. While it is true that the Milky Way has never collided with a galaxy as massive as Andromeda in the past the evidence of collisions between collisions of the Milky Way with smaller dwarf galaxies is growing.

Spiral galaxies like that of the Milky Way, produce new generations of stars for as long as they contain large molecular clouds of interstellar hydrogen inside the spiral arms. The spiral galaxies of the elliptical type are mostly without this gas which means that they do not produce many new stars. The quantity of material for star formation is limited; after stars have converted the hydrogen into more heavy elements the new star formation will cease.

The current period in the development of stars is anticipated to last for a period of one hundred billion years and it is expected that the "stellar time" will end at around ten trillion or 1100 trillion years (10^{13}-10^{14} years) and the smallestand longest-lived stars of our universe, such as tiny red dwarfs, start to fade. After the stellar age galaxy clusters will consist of compact objectssuch as Brown dwarfs, White dwarfs which have cooled or are cool ("black dwarfs") neutron stars and black holes. As the result of gravitational relaxation the stars will sink into central supermassive dark holes or be thrown into intergalactic space because of collisions.

Galaxies are magnetized with magnetic fields that are their own. They're powerful enough to be dynamically significant. They drive masses into the center of galaxies. They alter the development of spiral arms and also affect the gas's rotation within the regions that are in outer galaxies. Magnetic fields facilitate the transfer of angular momentum that is required for the disintegration of gas clouds and the creation of stars.

The standard strength for equipartition for spiral galaxies is around 10 million mG (microGauss) (or 1 nanoT (nanoTesla). To give an example, magnetic field on Earth is around 0.3 G (Gauss or 30 millimeters (microTesla). Galaxies with radio-faint spectra like M 31 and M 33 which are our Milky Way's closest neighbors are characterized by weaker fields (about 5 million volts) and gas-rich galaxies, with high star formation rates, such as M 51, M 83 and NGC 6946, have 15mG on average. For the prominent spiral arms the field strength could be as high as 25 mG in areas in which dust and cold gas are as well. The most powerful fields of total equipartition (50-100 mg) were discovered in starburst galaxies. For instance, those in M 82 and the Antennae and also in regions of nuclear starburst, such as in the central regions of NGC 1097 as well as the other barred galaxies.

Infrared-lit galaxies that are luminous or LIRGs are galaxies having luminosities. This is the measurement of the electromagnetic power output at or above 1011 L (solar luminosities). Most of the energy is generated by massive number of young stars that heat dust around them and then emits the heat in the infrared. A luminosity that is sufficient to qualify as an LIRG requires a star-forming rate of at least 18 million yr-1. Infrared galaxy clusters that are ultra-luminous (ULIRGs) can be at least ten times brighter than normal galaxies and can produce stars at speeds of more than 180 M yr-1. A large number of LIRGs are also emitting radiation from AGNs. Infrared galaxies produce much more infrared energy spectrum than at other wavelengths with the highest emission usually occurring at wavelengths between 60 and 100 microns. LIRGs are rare within the local Universe but they were frequent in the past when the Universe was smaller.

A part of the visible galaxies can be considered active galaxies when the galaxy has the galactic nucleus that is active (AGN). A significant proportion of the energy generated by the galaxy is generated by the galactic nucleus that is active in place of dust, stars and interstellar medium that comprise the galaxy. There are a variety of classifications and names for AGNs and those that fall that are in the lower regions of luminosity are referred to as Seyfert galaxies (discuss in a later article) and ones with luminosities higher than those of that of the galaxy host are referred to as quasi-stellar objects (also known as quasars). AGNs emit radiation across the electromagnetic spectrum, from radio wavelengths up to X-rays although certain portions of the radiation could be absorption by gas or dust in the vicinity of the AGN as well as their host galaxy.

The most common model of an active galactic nucleus built on an accretion disc that is formed around a massive black hole (SMBH) located in the center part of galaxy. The radiation coming from an active galactic nucleus comes from the gravitational energy produced by matter when it is pushed towards the black hole through the disc. The brightness of an AGN is determined by how massive the SMBH and the speed that matter falls on it. In approximately 10 percent of these galaxies two diametrically opposing pairs of jets with high energy ejects particles from the core of the galaxy at speeds that are that are close to the speed of light. The mechanism behind the production of the jets is not fully known.

Stars are formed within galaxies through a supply of cold gas which transforms into massive molecular clouds. Certain galaxies are observed to create stars at a rapid speed, referred to as the starburst. Should they continue to continue doing this it means they'll consume the gas reserves they have in a shorter time than the life span that the galaxies have. Therefore, starburst activity typically lasts just a few million years, which is a small time span in the evolution of the galaxy. Starburst galaxies were more prevalent in the beginning of the universe. They are, to this day, contribute around 15% of the total production of stars.

Starburst galaxies are distinguished by gas-rich dust along with the appearance new stars which include massive stars that are able to ionize the clouds around them to form H II areas. Massive stars cause supernova explosions. They result in the expansion of remnants which interact strongly with the gas surrounding them. The explosions cause the chain reaction of star-building that propagates throughout all of the region that is gaseous. Only after the gas available is depleted or dispersed, does the starbursts stop.

Starbursts are typically associated with merging or interfering galaxies. One example of the starburst-forming interplay is M82 which had an intimate encounter with the bigger M81. In irregular galaxies, there are often knots that are spaced out of starburst activity.

The Blazars have been identified as an active galaxy that has relativistic jets that are facing towards Earth. Radio-emitting galaxies emit radio frequencies due to relativistic jets. A unifying model of these kinds of galaxies that are active explains their distinct characteristics based on the observational angle of the observer.

Seyfert galaxies form one of the two main groupings of active galaxies alongside the quasars. They possess quark-like nuclei (very luminescent large and distant sources of radiation) with extremely high luminosities on their surfaces, but unlike other quasars their host galaxies are easily discernible. Seyfert galaxies comprise approximately 10% of galaxies. In visible light, the majority of Seyfert galaxies appear like normal spiral galaxies. However when studied at other wavelengths, the brightness of their central regions is similar to the luminosity of entire galaxies that are the size of the Milky Way.

Quasars, or quasi-stellar radio source are the most active and distant components of galactic nuclei that are active. Quasars are extremely bright and were initially identified as sources with high redshift of electromagnetic energy. This includes visible and radio waves which were believed to be like stars, rather than distant sources comparable to galaxies. Their luminosity is 100 times greater than that of the Milky Way.

Despite the popularity of huge spiral and elliptical galaxies the majority of galaxies are dwarf galaxies. They are comparatively tiny in comparison to other galactic formations. They are approximately one hundredth that of Milky Way, containing only one billion stars. The ultra-compact dwarf galaxy has recently been discovered and are 100 parsecs in size. [81]

Many dwarf galaxies could orbit one larger galaxy. The Milky Way has at least twelve satellites like this and an estimated 300-500 still being discovered. [82It is believed that there are 82 Dwarf galaxies could additionally be classified ellipticals spiral or irregular. Small dwarf ellipticals show no resemblance at all to large ellipticals they are usually described as dwarf spheroidal galaxies instead.

A investigation on an analysis of Milky Way neighbors found that for all dwarf galaxies the central mass of the galaxy is about 10.3 million solar masses no matter if the galaxy contains millions or thousands of stars. This is the reason for the hypothesis that galaxies are mostly formed by dark matter. It is also suggested the size of the minimum may be a sign of a warm dark matter, which is ineffective at co-evolution with gravity at a lower scale.

Galaxies are available in three primary types: spirals, ellipticals and irregulars. Let's talk about them.

1. Elliptical -Elliptical - Hubble classification system evaluates galaxies with elliptical shapes by their ellipticity. It ranges from E0 almost circular, to E7 which is extremely stretched. They are ellipsoidal in profile, which gives an elliptical look regardless of the angle from which they are observed. Their appearance suggests little structure and generally have minimal interstellar material. Therefore, they contain a small percentage of open clusters as well as an eroding rate of the formation of new stars. They are instead heavily dominated by older, more

advanced stars which orbit the center of gravity at random angles. They have low levels of heavy elements due to the fact that star formation ceases following the initial explosion. In this way, they share some resemblance to smaller clusters that are globular.

The biggest galaxies are ellipticals. Many elliptical galaxies are thought to have formed due to the interaction between galaxies, leading to a collision and merging. They can expand to massive dimensions (compared with spiral galaxies for instance) and massive spiral galaxies are usually located near the center of galaxy clusters with large sizes.

2. Shell galaxy Shell galaxy is a kind of elliptical galaxy in which the stars within the galaxy's halo have been placed within concentric shells. About one-tenth of the elliptical galaxy have a shell-like shape, which is not seen within spiral galaxies. These shell-like structures are believed to be formed when a bigger galaxy absorbs a smaller galaxy. When the two centers of galaxy are near, the centers begin to

rotate around a central point. This creates waves of gravitational energy that create the stars' shells, like ripples that spread on water. For instance, galaxy NGC 3923 is home to more than twenty shells.

3. Spirals are Spiral galaxies look like spiraling pinwheels. While the stars and the other visible elements in these galaxies are mostly on an uniplane, the majority of the mass in spiral galaxies is the form of a spherical-shaped halo of dark matter that extends over the visible component as shown by the universal rotation curve.

Spiral galaxies comprise the rotating disk of stars and an interstellar medium, as well as the central bulge, which is typically made up of older stars. A galaxy with weakly defined arms is often called an undefined spiral galaxy contrary to the larger design spiral galaxy with clearly defined and clearly defined spiral arms. The speed at which a galaxy spins is thought to be linked to the disc's surface because some spiral galaxies have massive bulges, whereas others are small and compact.

In spiral galaxies the spiral arms exhibit the appearance of logarithmic spirals. It is which can be theoretically proven to arise from a disturbance within an evenly moving mass of stars. As with the stars, spiral arms revolve around their center, however they do it with a an angular rate that is constant. These arms can be believed to be regions of high-density matter or "density waves". When stars traverse in an arms, speed of light of the stellar system is altered due to the gravity force that is associated with the dense region. (The velocity is restored to normal once the stars have left to the other end of the arms.) The effect is similar to an "wave" that slows down that travel through a road full of moving vehicles. The arms can be seen because the density is high, which facilitates star formationand they contain a lot of young and bright stars.

4. Barred spiral galaxy Most spiral galaxies, like our galaxy, our own Milky Way galaxy,

have an elongated, bar-shaped line consisting of stars, which extends towards the opposite side of the core and then joins to form the structure of spiral arms. [69] According to Hubble's Hubble classification system, they are identified with an SB which is and then a lower case letters (a, b or) that indicates the shape that the arms are spiral (in similar fashion to the classification of spiral galaxies that are normal). Bars are believed to be non-permanent structures which may arise as a result of a dense wave radiating outwards from the center, or because of a tidal connection with other galaxies. A lot of barred spiral galaxies are active, perhaps due to gas flowing into the core via the arms.

Recently, researchers have identified super-luminous spirals, also known as galaxies. They are massive, with an upwards diameter of 437,000 light-years (compared with the Milky Way's 100,000 light years diameter). With an estimated around 340 billion solar masses they emit a large amount of mid-infrared as well as ultraviolet light. They are believed to have an enhanced star formation rate of around 30 times more quickly than the Milky Way. The majority of galaxies are gravitationally bound to other galaxies. They form a hierarchical fractal-like arrangement of clustered structures with the most compact of these associations being referred to as groups. Groups of galaxies are the most frequent type of galactic cluster and they contain the large portion of galaxies (as well as the bulk part of the mass baryonic) within the universe. To remain bound by gravitational force to a group, each galaxy must be at a low velocity that it is unable to stop from leaving (Virial theorem (explain it in a moment)). If there is not enough energetic kinetic energy the group might be transformed into a smaller amount of galaxies due to mergers.

The clusters of galaxies comprise many thousands to hundreds of galaxies that are bound with gravity. Galaxies clusters are usually controlled by one giant spiral galaxy, also often referred to as the most bright cluster galaxy. It gradually eliminates its satellite galaxies, and then adds their masses to its own. Superclusters are composed of tens of thousands of galaxies. They are located in groups, clusters and occasionally in individual. On the scale of superclusters the galaxies are laid out in filaments and sheets that surround vast empty space. Beyond this the universe appears be identical in every direction.

It is believed that the Milky Way galaxy is a part of an organization known as"the Local Group, a relatively tiny group of galaxies which is approximately one megaparsec. It is believed that the Milky Way and the Andromeda Galaxy are the two most bright galaxies in the group. Many of the other galaxies in the group are dwarf counterparts to the two. This Local Group itself is a part of a cloud-like structure inside the Virgo Supercluster that is a vast extended structure that includes galaxies in clusters and groups focused around the Virgo Cluster. Additionally, the Virgo Supercluster is element of the Pisces-Cetus

Complex Supercluster, an enormous galaxy filament.

The highest radiation of the majority of stars is within the spectrum of visible light, which is why the study of galaxies formed by the stars is an important aspect in optical astronomy. It's also a great part of the spectrum for observation of the ionized H II regions, as well as for studying how dusty arms are distributed.

Ultraviolet and Xray telescopes can observe galactic phenomena that are extremely energetic. The appearance of ultraviolet flares is often observed when the star of distant galaxies is torn away from the tidal force of the near black hole. Galactic clusters with hot gases could be detected by the X-rays. The existence of massive black holes in the central regions of galaxies has been confirmed by the X-ray astronomy.

I didn't really know much about the virial theorem. I suggest you search for it to get a better understanding. I'm able to give you an some idea of what the virial theorem actually is? According to

Wikipedia the virial theorem is an equation general enough to connect an average of time over the entire energetic energy of a steady system of discrete particles connected by potential forces to the total energy potential of the system.

For those who wish to learn more about this theory, there are more details in relation to it.

The importance of the virial theory is that it permits the average total energy of kinetics to be calculated for extremely complex systems that do not have the exact solution, like the ones that are that are studied in statistical mechanics. the average total kinetic energy is correlated to temperature using an equipartition theory. The virial theorem doesn't rely on the concept of temperature, and is valid in systems that are not in equilibrium thermally.

As you are aware, I am a 17 years old and a physics fanatic. In addition, I'm learning various new skills as I write this book.

Chapter 8: The Planets

My this planet is home to me. You will argue that we won't be able to live on every planet , so what define planet is home? So I will tell them the same as you wouldn't knock on anyone's door only to begin living in their house, it is not a suitable place to live in any home, but you are able to remain in your house and the similar humans could remain in their home until they find a suitable home or, in other words, a planet that is habitable.

An astronomical object that orbits the sun or a stellar remnant that is huge enough to be round by gravity, but not large enough to trigger thermonuclear thermic fusion.

It is not understood exactly the process by which planets form. Most theories suggest that they formed through the collapse of a nebula to the formation of a thin disc made of dust and gas. Protostars form at its core, and is which is surrounded by a protoplanetary disk rotating. Through the process of accretion (a process that involves sticky collisions) dust particles within the disk gradually build up mass and grow into ever-larger bodies. Small concentrations of mass, called planetesimals are formed they accelerate the process of accretion by attracting additional materials through their gravitational attraction. These concentrations grow ever larger until they fall backwards due to gravity and form protoplanets.After the planet has reached the size of Mars in mass, it starts to build up an expanded atmosphere, which increases the rate of capture for the planetesimals due to atmospheric drag. Based on the time of accretion of gas and solids the formation of a giant planet, an ice giant, or even a terrestrial planet could be formed.

If the protostar is large enough so that it is able to form a star the remaining disk is taken away from its insides outward through photoevaporation, solar wind, the Poynting-Robertson drag along with other interfering effects. In the future, there could be a variety of protoplanets orbiting one another or the star However, in time, they will all collide to create a larger planet, or release material for smaller planets or protoplanets to absorb. The objects that are large enough to absorb the majority of material in their orbital surroundings and eventually become planets. Protoplanets that aren't involved in collisions can become satellites of planets via an act of gravitational capture. They may also remain in belts with other objects and evolve into small bodies or dwarf planets.

The energy-driven impacts of smaller planetesimals (as with radioactive decay) can heat the planet's expanding surface, which causes it to either partially or completely melt. The planet's interior starts to separate in mass, forming an increasingly dense core. Smaller terrestrial planets lose the majority of their atmospheres due to this process of accretion. However, gases that are lost can be replaced through outgassing from the mantle as well as after the subsequent collision of comets. (Smaller planets are likely to lose any atmosphere they acquire via various escape mechanisms.)

149

The discovery and the observation of planetary systems surrounding stars that are not the Sun and other stars, it is now possible to expand, modify or even completely replace this model. The level of metallicity - an scientific term used to describe the amount of chemical elements having an atomic number higher than 2 (helium)--is currently believed to be a determinant of the probability that a star will possess planets. This means that a metal-rich , population I star could have a larger planet system than a less metal-rich populations II stars.

The solar system's planets is divided into groups based on their composition...

* Terrestrials: Planets that are like Earth with bodies that are mostly comprised of rock: Mercury, Venus, Earth and Mars. With 0.055 mass of Earth, Mercury is among the most compact terrestrial planets (and the tiniest planet) within the Solar System. Earth is the biggest terrestrial planet.

* Giant Planets massive planets that are significantly bigger than terrestrials. Jupiter, Saturn, Uranus and Neptune.

* Gas giant Jupiter as well as Saturn are massive planets that are primarily made of helium and hydrogen. They are the largest planets within the Solar System. Jupiter at 318 Earth masses is the biggest world in the Solar System, and Saturn is one third of the size with 95 Earth masses.

* The giants of ice: Uranus and Neptune are mostly composed of low boiling-point substances like methane, water and ammonia, surrounded by dense atmospheres of hydrogen and Helium. They have a considerably lower mass than gaz giants (only 14 and 17 Earth masses).

The first time that we were able to confirm the existence of an extrasolar world orbiting an ordinary main-sequence star took place on October 6, 1995 the day Michel Mayor and Didier Queloz from the University of Geneva announced the discovery of an exoplanet in the vicinity of 51 Pegasi. From that point up to Kepler mission, the most well-known other solar planets are gas giants similar in size to Jupiter or even larger, as they were more easily observed. The catalogue of Kepler candidate planets consists mainly of planets that are the dimensions of Neptune or smaller, and down to smaller Mercury.

There are exoplanets more close to their star's parent than any other planet in our Solar System is to the Sun as well as there are exoplanets further away from their parent star. Mercury is the closest planet with respect to Sun at 0.4 an AU, requires about 88 days to complete an orbit however the shortest orbits of exoplanets are just about an hour. The Kepler-11 System (other sun system) contains five planets with smaller orbits than Mercury's, all of them bigger than Mercury. Neptune is located 30 AU away distant from the Sun and takes 165 years for its be in orbit, but there are exoplanets hundreds of AU away from their star and require more than 1,000 years to travel around.

A planetary mass object (PMO) or planemo or planetary body , is an celestial object that has masses that fall within the an actual planetas being large enough to reach the state of equilibrium in hydrostatics (to be rounded by its gravity) however not sufficient to support core fusion, as the star. According to the definition that is, all planets are planetary mass objects, however the intent of the phrase is to describe objects that don't conform to the typical definition of an actual planet. This includes dwarf planets that are rounded by their own gravity, but are not big enough to be able to clear their orbits as well as planetary-mass moons and planemos floating free, which could be expelled from a system similar to Rogue planets (discuss in a later post) or were formed due to cloud-collapse rather than accretion , as in sub-brown dwarfs (discuss in the future).

An astronomical dwarf is an astronomical mass object that isn't truly a planet nor a satellite that is natural; it is directly orbiting the star and is large enough for gravity to force to a hydrostatically equivalent shape (usually the Spheroid) however, it has not removed the surrounding area of other objects within its orbit. A planetary scientist, also known as New Horizons principal investigator Alan Stern who coined the term "dwarf planet" has stated that it is not necessary to consider the location of the object and that only the geophysical characteristics are to be considered and that dwarf planets are an alternative to planets.

A number of computer simulations of the stellar and the formation of planetary systems have suggested that certain objects with planetary mass could be ejected from interstellar space. They are often referred to as"rogue planets.

Stars are formed by the collapsing of clouds of gas however smaller objects may be formed by the collapse of cloud. The planetary mass objects that are formed this way are often called sub-brown-dwarfs. Sub-brown dwarfs could be floating like OTS 44, or they may orbit larger objects like 2MASS J04414489+2301513.

In close binary systems of stars, one of the stars could lose weight to a heavier counterpart. Pulsars that are powered by accretion could cause mass loss. The star that is shrinking could transform into a planetary mass object. One example of this is a Jupiter-mass Object that orbits the Pulsar PSR J1719-1438. The shrunken white dwarfs could transform into a helium-rich planet or carbon-rich planet.

One of the most important characteristics of planets is their magnetic moments that can lead to magnetospheres. If there is a magnetosphere, it suggests that the planet is active geologically. Also, magnetic planets are surrounded by electrically conducting materials within their interiors that create magnetic fields. These fields dramatically alter the interactions between the solar wind and the planet. A planet that is magnetically influenced creates a hole in the solar wind that surrounds it, referred to as the magnetosphere. This is a space that is inaccessible to the wind. The magnetosphere may be larger than the planet. However, non-magnetized worlds are characterized by small magnetospheres that are created through the interaction of the ionosphere in conjunction with solar winds that are unable to protect the planet.

Out of the 8 planets of the Solar System, only Venus and Mars do not have magnetic fields. Additionally Ganymede, the moon Jupiter Ganymede also has one. Of all the planets that are magnetized, the magnetic field on Mercury has the lowest strength and it is unable to block from the sun's wind. The magnetic field of Ganymede is many times stronger, and Jupiter's field is the strongest in the Solar System (so strong in the sense that it could pose an extremely health risk for the future astronauts who will be able to visit the moons). The fields that magnetically surround other planets of the giant are similar to the strength of Earth however their magnetism moments are much greater. These magnetic fields on Uranus as well as Neptune are significantly tilted with respect to the axis of rotation and are moved away from the center of the planet.

A group of astronomers from Hawaii discovered an extrasolar planet in the vicinity of the star HD 179949. The planet was able to create sunspots in the surface the star it's parent. The team speculated that the planet's magneticosphere is transferring energy to the surface of the star and thereby increasing its already-high temperature of 7,760 degrees Celsius by 400 degrees Celsius.

Each planet was born in a fluid state and in the early stages of formation, the heavier, denser materials fell to the centre while the lighter material remained on the surface. Every planet has a distinct interior made up of a thick planet's core that is enclosed by a mantle that is or was fluid. Terrestrial planets are enclosed by hard crusts. However, on the giant planets, the mantle blends in with the cloud layers above. The terrestrial planets are home to cores of elements like nickel and iron, as well as mantles made of silicates. Jupiter as well as Saturn believe to contain cores made of metal and rocks, which are that are covered by mantles of hydrogen from metallic sources. Uranus and Neptune which are both smaller, are rocky in their cores that are surrounded by mantles of ammonia, water, methane, and other Ices. The fluid motion in the cores of these planets produces a geodynamo which creates magnetic fields.

Each of the Solar System planets, with the exception of Mercury have massive atmospheres as their gravity is strong enough to hold atmospheric gases near to its surface. The giant planets that are the largest big enough to hold significant amounts of the gaseous hydrogen and helium, which are light in their weight while the smaller planets are able to release the gases to space. The atmosphere of Earth differs from other planets due to the diverse life-forms that have occurred on Earth have created the free oxygen molecules.

Planetary atmospheres are affected by the varying insolation or internal energy, leading to the formation of dynamic weather systems such as hurricanes, (on Earth), planet-wide dust storms (on Mars), a greater-than-Earth-sized anticyclone on Jupiter (called the Great Red Spot), and holes in the atmosphere (on Neptune). There is at least one extrasolar planet, HD 189733 b, is believed to possess this type of weather system like it's counterpart, the Great Red Spot but twice as big.

Hot Jupiters, as a result of their extremely close proximity towards their star hosts have been found to lose their atmospheres to space as a result of stellar radiation, just like comet tails. The planets might have massive variations in temperature between their night and day sides, which produce supersonic wind but the night and day sides of HD 189733b seem to have similar temperatures, which suggests the atmosphere of the planet is able to redistribute the energy of the star around the planet.

A number of dwarf planets or planets within the Solar System (such as Neptune and Pluto) have orbital times of resonance either with other smaller objects (this is also the case for satellites). The exceptions are Mercury as well as Venus possess natural satellites commonly referred to as "moons". Earth has one moon, Mars also has 2 moons, while the massive planets have a variety of moons within complex planetary systems. Numerous moons of the giant planets feature features that are similar to those found on the terrestrial planets as well as dwarf planets. A few are being studied as possible habitats for living things (especially Europa).

Four giant planets connected by rings that vary in size and complex. They are made up of a majority of particulate matter or dust and can also host tiny moonlets that's gravity shape and sustains their shape. Although the origins of the planetary rings aren't fully understood however, they are believed to have been caused by natural satellites that fell beneath their planet's

Roche limit and were ripped apart by the force of tidal forces.

I will mention some other types of planets to help understand planets better.

A binary system in which two objects of mass planetary orbit each other have an orbital axis that is external to both . Two planets with mass which orbit one another is known as Double planet.

Minor planets are meteorological object with a close orbit with the Sun which is not considered a planet or designated as a comet.

Mesoplanets are objects of planetary mass with sizes that are smaller than Mercury but bigger than Ceres.

Chapter 9: Moon

Moon is the sole natural satellite that is our own blue planet Earth the only other place on which humans have entered. Before we discover the most amazing facts about our moon, it is important to know all about the natural satellite, and satellites. Let's get started.

Natural satellites, also known as a moon, is in the most popular sense an astronomical object which orbits a star or minor planet.

Natural satellites orbiting close to the earth in an upward trajectory (Retrograde motion within astronomy refers to generally described as the motion of an orbit or rotation of objects in the direction that is opposite to the direction of rotation of its primaryobject, which is the central object.) Uninclined circular orbits (regular satellites) are believed to have originated out of the collapsed area in the protoplanetary disk, which formed its first. A majority important natural satellites in the Solar System have regular orbits however, the majority of small satellites are in irregular orbits. The Moon and perhaps Charon (largest five-known native satellite from pluto) are a few exceptions to the big bodies, as they are believed to have been formed by collisions between two huge proto-planetary objects. The material that could have been put within orbit around this central object is thought to have reaccreted, forming some natural satellites that orbit. Contrary to planet-sized bodies the moons of asteroid are believed to be formed by this process. Triton (largest identified natural satellites of the planet Neptune) is an additional exception though it is massive and in an orbit that is close and circular it's motion is retrograde , and it is believed to be a dwarf captured planet.

Most moons of regular size (natural satellites that follow very close orbits with a low inclined and eccentricity) within the Solar System are tidally locked to their primary orbits which means that the opposite part of the original satellite is in front of its planet. The only exception to this is the Saturn's satellite of natural origin Hyperion that rotates in a chaotic manner because of the gravitational impact of Titan.

Contrary to this, the outside natural satellites orbiting the globes (irregular satellites) are too far to be locked. For instance Jupiter's Himalia as well as Saturn's Phoebe along with Neptune's Nereid have rotation times between ten and 10 hours, whereas their orbital times vary from hundreds of days.

Discovering 243 Ida's satellite natural satellite Dactyl in the 1990s, confirmed that certain asteroids do have satellites from nature; indeed the there are 87 Sylvia has two satellites. Certain, like 90 Antiope are double asteroids that have two similar-sized components.

Of the 19 natural satellites within the Solar System that are sufficient in size to have passed into equilibrium with the hydrosphere, a few are still active in the geological world today. Io is by far the largest active volcano within the Solar System, while Europa, Enceladus, Titan and Triton show indications of ongoing tectonic activity as well as cryovolcanism.

In the three instances the activity in the earth is driven by the tidal warming that results from the eccentric orbits that lie near their giant-planet primaries. (This mechanism may have also worked in Triton during the time of its formation before it was circularized.) Other natural satellites like the Earth's Moon, Ganymede, Tethys and Miranda are evidence of geological activity in the past caused by energy sources like that of the decomposition process of their radioisotopes from the beginning and greater orbital eccentricities in the past (due in certain cases to orbital resonances in the past) or the differentiation or the freezing in their inner regions. Enceladus and Triton both possess active features that resemble geysers but in the case Triton solar heating is believed to generate the energy. Titan as well as Triton have large atmospheres. Titan additionally has lakes of hydrocarbons. Also, Io as well as Callisto are surrounded by atmospheres even though they're extremely thin. Four of the biggest terrestrial satellites that are natural, Europa, Ganymede, Callisto and Titan are believed to contain subsurface oceans of liquid water. The smaller Enceladus could be home to the subsurface water is localized and liquid.

A quasi-satellite is a spacecraft with a specific co-orbital configuration with the planet, where the object remains close to the planet during multiple orbital periods.

A subsatellite, also referred to as a moonmoon, submoon, or moonmoon is a natural or artificial satellite that orbits around a naturally-occurring satellite i.e. it is a "moon of moon".

We have studied enough about the natural satellite, I told you that after we have learned about the natural satellites we'll begin to understand our moon. We'll start by studying the process of moon formation and let's continue...

The Moon was created 4.51 billion years ago, or perhaps more than 100 million years before, or 50 million years before the beginning of the Solar System, as research published in the year 2019 suggests. Numerous theories of formation have been suggested for the Moon, such as the separation that separates moon from the earth's surface, or fission of Moon away from the Earth's crust by centrifugal force (which requires too large an initial rate of rotation for Earth) and the gravity-based capture of a preformed Moon[35[35] (which will require an impractically expanded the atmosphere on Earth to absorb the energy from the Moon's passing Moon) and formation that brought Earth together with Earth and the Moon as a whole in the initial disk of accretion (which is not the reason for the depletion of the metals found in Moon's accretion disk). Moon). These theories also do not explain the high acceleration of the Moon-Earth system.

The development of the Moon and a tour around the Moon

The prevalent hypothesis is that Earth-Moon-system formed due to a huge collision of a Mars-sized object (named the Theia) along with proto-Earth. The impact was able to blast material into Earth's orbit, and the material accreted to form the Moon.

This theory best describes the evidence. In the 18 months preceding the October 1984 conference on the lunar origins Bill Hartmann, Roger Phillips along with Jeff Taylor challenged fellow lunar scientists.

Massive impacts are believed to have been frequent in the beginning of the Solar System. Computer simulations of huge impacts have yielded results that match the size that the moon's core carries as well as the angular momentum of Earth-Moon system. The simulations also reveal that the vast majority of the Moon was derived from the impactor instead of the proto-Earth. However, recent simulations suggest a higher percentage of the Moon is derived directly from Earth's proto. Other bodies within the Solar System such as Mars and Vesta are, according to meteorites they have, distinct oxygen and tungsten isotopic compositions when compared to Earth. But, Earth and the Moon are nearly identical in their isotopic compositions. The isotopic balance of the

Earth-Moon axis could be explained through the mix of the post-impact material that created the two, but this topic is disputed.

The impact released lots of energy, and the released material was re-accreted to the earth-Moon system. It would have also melted the exterior shell Earth and subsequently formed an ocean of magma. Similar to the newly created Moon could have also been affected, and even had its own lunar magma sea and its depth ranges as ranging from 500 kilometers (300 miles) to 1,737 km (1,079 miles).

The theory of the giant impact is the basis for a variety of research however, there are some issues that remain unanswered. The majority relate to the moon's composition.

In 2001, a group from the Carnegie Institute of Washington reported the most precise measurements of isotopic fingerprints on lunar rocks. The rocks of the Apollo program were found to have the same isotopic signatures that rocks from Earth and were different from every other body within The Solar System. This discovery was surprising, since the majority of the material that made up the Moon was believed to have come directly from Theia as it was revealed by the NASA in 2007 that there is less than percent possibility there was a chance that Theia and Earth have identical isotopic signatures. Another Apollo lunar samples showed similar titanium composition as Earth and Earth, which is in contradiction to what we would expect in the event that The Moon was formed further away from Earth or was in fact derived from Theia. These differences can be explained through variations of the theory of giant-impact.

Liquid water can't remain in the moon's surface. If exposed to sunlight the water rapidly decomposes in an process called photodissociation before being lost to space. Since during the decade of the 1960s, researchers have proposed that water ice might be formed by collisions with comets or perhaps formed by the reaction between moon rocks that are oxygen-rich, as well as the solar wind's hydrogen leaving water traces that could remain in the cold, permanently shadowed in craters located at the poles of the Moon.

This year's Chandrayaan-1 spacecraft has verified the presence of water ice on the surface with the onboard Moon Mineralogy Mapper. The spectrometer detected absorption lines, which are typical of the hydroxyl in the reflected light and provided evidence of huge amounts of water ice in the moon's surface. The spacecraft revealed that the concentrations could exceed 1000 ppm. Utilizing the mapper's reflectance spectrum indirect illumination of regions that were shadowed confirmed the presence of the presence of water ice in 20deg latitude between the poles during the year 2018.

Analyzing the results that the Moon Mineralogy Mapper (M3) released in August 2018 provided as the very first "definitive scientific evidence" for the presence of water-ice in the moon's surface. The results revealed distinct reflection signatures of water-ice as opposed in other reflective materials. Ice deposits were observed at the North as well as the South poles, however it is most abundant in the South as the water remains in obscured crevices and craters, which allows it to remain as an ice layer on the surface as they are protected from the sun.

In October of 2020 scientists reported finding molecular water on the sunlight-lit moon's surface Moon by a variety of spacecraft independent of each other which include that of the Stratospheric Observatory for Infrared Astronomy

The gravitational field on the Moon is measured by monitoring the Doppler shift in radio signals that are emitted by spacecraft that orbit. The most prominent lunar gravity features are mascons. They are large positive gravitational anomalies that are associated with certain of the massive impact basins, which are primarily due to the massive basaltic lava flows in the mare which fill these basins. These anomalies significantly affect the orbit of spacecraft around the Moon. There are a few puzzles to solve that lava flows on their own do not explain the whole gravitational signature, and a few macons exist, but are not connected to mare volcanism.

It is believed that the Moon also has an outer magnetic field typically lower than 0.2 nanoteslas, which is just one-hundredth of of Earth. The Moon has not yet an overall dipolar magnetic field , and has crustal magnetization , which was likely acquired at a time in the time when a dynamo still operating. In the beginning of its history, about 4 billion years prior, the magnetic field was comparable to Earth now. The first dynamo field was wiped out around 1 billion years ago, when the lunar core had fully crystallized. Theoretically, some remaining magnetization might result from temporary magnetic fields created during large-scale impacts, due to the growth in plasma cloud. These clouds are formed by large impacts within an atmosphere with a magnetic field. This is confirmed by the position of the biggest crustal magnetizations, which are located close to the antipodes of the massive impact basins.

It is believed that the Moon features an atmosphere that's so fragile that it is almost empty, with an overall mass that is less than 10 tons (9.8 long tons; 11. shorter tons). The pressure on the surface of this tiny body is about 3x10 to atm (0.3 nPa) and varies according to moonlight. The sources of this mass include outgassing and sputtering. These are caused by blasting of lunar soils by solar wind particles. Some elements detected comprise potassium and sodium generated through Sputtering (also discovered within the atmospheric layers that surround Mercury as well as Io) along with neon and helium-4 from the solar wind and the argon-40, radon-222 and polonium-210that outgassed following their formation by radioactive decay in the mantle and crust. The absence of as neutral substances (atoms as well as molecules) like nitrogen, oxygen carbon, hydrogen, and magnesium found within the regolith (Regolith is an enveloping composed of loose, unconsolidated uneven superficial deposits that cover rock bodies such as moon, mars and earth) It is unclear why they are absent. Water vapor has been observed by Chandrayaan-1 and observed to fluctuate in latitude, reaching the highest concentration occurring between 60 and 70 degrees. It could be caused by the sublimation of water ice inside the regolith. These gases can either escape into the regolith due to the gravity of the Moon or escape to space via the pressure of solar radiation or, when they become ionized due to disappearing due to magnetic field of the solar wind.

The Moon is spinning around it's own orbital axis. The reason for this is lock-up of the tides synchronous with its orbital time around Earth.

The duration of the rotation is dependent on the reference frame. There are sidereal period of rotation (or sidereal day with respect with the constellations) as well as synodic rotation times (or synodic day with respect with in relation to the Sun). A lunar day is a synodic day.

Due to the locked tidal rotation the synodic and sidereal orbital times correspond with the synodic (27.3 Earth days) and synodic (29.5 Earth days) orbital times.

The Moon completes its circle around Earth in relation to fixed stars around every 27.3 days[h[h] (its time of sidereal motion). But, since Earth is moving in relation to its Sun simultaneously, it can take longer to allow the Moon to display the same phase as Earth approximately 29.5 days[i[i] (its synodic time). In contrast to other satellites on planets however, the Moon is located more closely to its ecliptic line as opposed to Earth's equatorial plane. This orbit of the Moon is affected through Earth and the Sun as well as Earth in various tiny, intricate and interconnected ways. For instance the lunar plane orbit slowly rotates each 18.61 days, this alters the lunar motion in other ways. The follow-on effects can be mathematically described in Cassini's laws. The Moon is a huge natural satellite in relation to Earth The Moon's diameter is greater than one quarter, while its volume is one-tenth of Earth's. It is the biggest moon of the Solar System relative to the size of the planet it orbits, however, Charon is larger than that of the small planet Pluto with a ratio of 1/9 Pluto's mass. Charon is the Earth and the moon's Barycentre (the barycenter represents the centre of mass for several bodies which orbit each other, and is the location around where the bodies rotate.) Their common mass center, is located 1,700 kilometers (1,100 miles) (about half of the Earth's diameter) below the surface of Earth.

It is believed that the Earth rotates about the barycentre at the Earth-Moon orbit at least once every sidereal month at 1/81 of that of the Moon approximately 12.5 metres (41 feet) every second. The motion of the Moon is set upon the larger rotation that revolves the Earth about the Sun at a rate of around 30 km (19 miles) each second.

The total size that is the Moon is slightly smaller than that from North as well as South America combined.

The gravitational attraction masses share with one another decreases in inverse proportion to the square of the distance that these masses are from one another. This is why the slightly higher attraction the Moon is attracted to the portion of the Earth close to the Moon in comparison to the portion of the Earth which is in opposition to the Moon which results in the tidal force. Tidal forces impact both oceans and the Earth's crust.

Although gravitation can cause acceleration and movement in the oceans that flow through the Earth Gravitational coupling between Moon with Earth's solid body is flexible and flexible. The result is a impact of Moon's impact on the Moon upon the Earth that creates an increase in the solid part of the Earth close to the Moon which acts as an energy source in opposition to the Earth's rotating. This "drains" the angular momentum as well as the kinetic energy of rotation from the Earth's rotating, slowing Earth's movement.

The angular momentum that is lost from the Earth is then carried to Moon through the process (confusingly called tidal speed) which elevates the Moon to a higher orbit, resulting in a lower speed of orbit around the Earth. The distance between Earth and the Moon is growing and the Earth's orbit is slowing down in response. Laser reflectors that were measured on Apollo's Apollo missions (lunar range experiments) have shown that the Moon's distance grows by 38 millimeters (1.5 in) each year (roughly the speed of the human fingernails growing).

The drag of the tidal current would persist until the rotation of Earth and the orbital duration of the Moon match, resulting in mutual tidal locking between two, and suspending both the Earth and the Moon over a single meridian. Sun will turn into an enormous red giant that will engulf the Earth-Moon system before this event. If this locking of the tides occurred, the rotation of the Earth will continue to slow because of the tides created from the Sun. As the days get more than the monthly and the Moon will then slowly move from east to west across the sky. The tides created by the Moon will then create the opposite effect as before it, and the Moon will move closer to Earth. The Moon would eventually get within of the Roche limits (the Roche limit, also known as Roche radius), which is the distance to the celestial body that the second celestial body that is held together by the force of its own gravity, would break apart because the first's tide forces are greater than the gravitational self-attraction of the second) and then break into rings.

The cumulative effect of stress accumulated by these forces causes moonquakes. Moonquakes are less frequent and less powerful than earthquakes however, they may last up to one hour - much longer than terrestrial earthquakes - as a result they do not have water to absorb the vibrations of seismic waves. Moonquakes are an unanticipated discovery made by seismometers that were placed upon the Moon by Apollo astronauts in 1969 and 1972.

The Moon was once spinning at a higher speed however, early in its development, the Moon's rotation slowed and it was tidally locked into this position because of frictional effects caused by the deformations of the tidal axis due to Earth. In time the energy from the its rotation Moon on its axis lost as heat until there was no movement of the Moon in relation to Earth. In 2016, scientists studying planetary science using data from the older NASA Lunar Prospector mission, discovered two hydrogen-rich areas (most likely once water ice) located on opposing sides of Moon. It is thought that these patches represent two poles that were on the Moon many billions of years ago, before it was locked tidally to Earth.

Since that the moon's orbital orbit Earth has a slope of approximately 5.145deg (5deg 9.0') to the orbit of Earth around the Sun The Sun's orbit is not a perfect circle, so eclipses cannot occur on every full or new moon. To be able to experience an eclipse to occur, it is necessary for the Moon must be located near the intersection of two planes of orbit. The frequency and frequency that occur during eclipses, of the Sun through the Moon and of the Moon through Earth can be described in the saros. It has an approximate period of 18 years.

Since the Moon continually obscures views of the half-degree circle of sky, the corresponding phenomenon of occultation takes place when an astronomical object or star is positioned in front of the Moon and becomes occulted, obscured from view. This is why the solar eclipse is an occultation of Sun. Since the Moon is relatively near to Earth the occultations of particular stars aren't visible across the entire planet or even simultaneously. Due to the phenomenon of precession (Precession is the change in the direction of the axis of rotation of the body that is rotating.) in the moon's orbit each year various stars are obscured.

Conclusion

Astronomy is a fascinating science that raises numerous questions we're still looking for answers to. For those who are not experts it opens a door into the universe around us, and the awe-inspiring beauty. It helps us be aware of how small we are to the vastness of the universe, and we are humbled by the beauty of nature in general. Astronomy is also a fantastic opportunity to have a constructive time with your friends. On my own I'd like to thank you for being able to complete this book and I'm hoping it is the beginning of your enthusiasm and love for exploring space.

9 781774 857120